Environmental Engineering

Series Editors: U. Förstner, R.J. Murphy, W.H. Rulkens

Springer

Berlin
Heidelberg
New York
Barcelona
Hong Kong
London
Milano
Paris
Tokyo

Alena Mudroch, M.Sc., was a research scientist and Chief of Sediment Remediation Project at the National Water Research Institute, Environment Canada, Burlington, Ontario. Her major research interests include the characterization of aquatic sediments, defining the role of the pathways, fate, and effects of sediment-associated contaminants in aquatic ecosystems, remediation of contaminated sediments, and effects of mining on aquatic environments. She is editor and co-author of six books dealing with the above subjects.

Ulrich Stottmeister studied chemistry at Leipzig University (Dr. rer. nat. 1968). From 1970 to 1990, he was scientific coworker and Head of the Department at the German Academy of Sciences, Institute of Biotechnology Leipzig. He received his degree in biotechnology (Dr. sc. nat.) in 1986. He was Assistant Professor at the University of Waterloo, Canada, in 1990. U. Stottmeister has been Head of the Department of Remediation Research of the UFZ Centre for Environmental Research Leipzig/ Halle since 1992. Since 1995 he has an appointment as Professor for Biotechnology at Leipzig University, Faculty of Chemistry and Mineralogy. He is a full member of the Saxonian Academy of Sciences. His main fields of work include: microbial overproduction, industrial wastewater treatment, and remediation research in disturbed landscapes. He has published approximately 100 scientific papers in journals and books and is co-editor of several books.

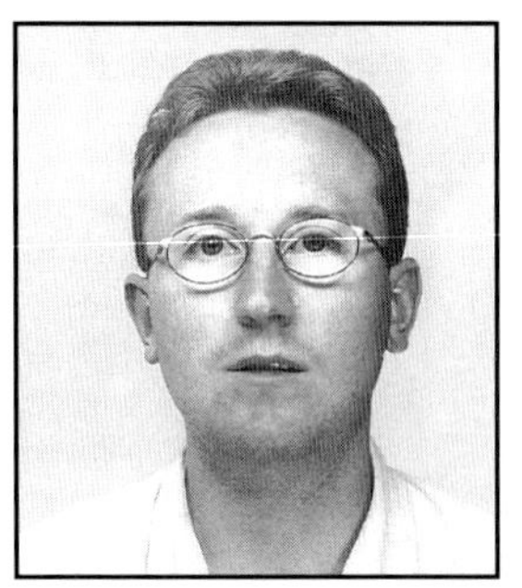

Christopher Kennedy is Assistant Professor in Environmental Engineering at the University of Toronto, where he teaches courses in Groundwater, Hydrology, Hydraulics and Statistics. His research interests include subsurface contamination/remediation, reclamation of surface-mined lands and the design of sustainable urban infrastructure. He holds degrees in Civil Engineering from the University of Waterloo, Canada, and Imperial College, London, UK.

Helmut Klapper, now senior scientist at the Institute for Water Research Magdeburg, began his research in 1956 on applied hydrobiology: biomonitoring for classification of rivers and lakes, biological processes in groundwater enrichment, control of water pollution. At the end of the 1960s, he turned to research and development of ecotechnologies to combat eutrophication, e.g. deep-water delivery, hypolimnion aeration, nutrient precipitation, sediment capping with calcite, etc. Actual fieldwork includes the new forming lakes from opencast lignite mining, some of which are sulfur acidic with pH 2—3.5. New approaches in water quality management focus on the microbial reversal of the acidification by desulfurization in anoxic parts of the ecosystems. Professor Klapper has published more than 200 papers and books.

Alena Mudroch · Ulrich Stottmeister
Christopher Kennedy · Helmut Klapper (Eds.)

Remediation of Abandoned Surface Coal Mining Sites

A NATO-Project

With 73 Figures and 22 Tables

Series Editors:
Prof. Dr. U. Förstner — Arbeitsbereich Umweltschutztechnik, Technische Universität Hamburg-Harburg, Eißendorfer Straße 40, 21073 Hamburg, Germany
Prof. Robert J. Murphy — Dept. of Civil and Environmental Engineering, University of South Florida, 4202 East Fowler Avenue, ENG 118, Tampa, Fl 33620, USA
Prof. Dr. ir. W.H. Rulkens — Wageningen Agricultural University, Dept. of Environmental Technology, Bomenweg 2, P.O. Box 8129, 6700 EV Wageningen, The Netherlands

Authors:
Dr. Alena Mudroch — 9 Barrie St., Dundas, L9H 4S6 Ontario, Canada, Department of Civil Engineering, University of Toronto, 35 St. George Street, M5S 1A4 Toronto, Ontario, Canada
Prof. Ulrich Stottmeister — Department of Remediation Research, UFZ Centre for Environmental Research, Permoserstr.15, 04318 Leipzig, Germany
Prof. Christopher Kennedy — 9 Barrie St., Dundas, L9H 4S6 Ontario, Canada, Department of Civil Engineering, University of Toronto, 35 St. George Street, M5S 1A4 Toronto, Ontario, Canada
Prof. Helmut Klapper — Schrotebogen 10, 39126 Magdeburg, Germany

ISBN 3-540-42539-X Springer-Verlag Berlin Heidelberg New York

Library of Congree Cataloging-in-Publication Data
Remediation of abandoned surface coal mining sites / A. Mudroch ... [et al.].
p. cm. -- (Environmental engineering) Includes bibliographical references and index
ISBN 3-540-42539-X
1. Coal mines and mining--Environmental aspects. 2. Strip mining--Environmental aspects.
3. Abandoned mined lands reclamation. I. Mudroch, Alena. II. Environmental engineering
TD195.C58 R45 2001 631.64--dc21 2001053285

Springer-Verlag Berlin Heidelberg New York
a member of BertelsmannSpringer Science+Business Media GmbH

httpp://www.springer.de

Printed in Germany

Typesetting: Dataconversion by authors
Cover-design: Struve & Partner, Heidelberg
Printed on acid-free paper SPIN: 10850635 30 / 3020 hu - 5 4 3 2 1 0 -

Executive Summary

Christopher Kennedy

University of Toronto, Department of Civil Engineering, 35 St. George Street, Toronto, Ontario, Canada M5S 1A4

Coal is a valuable resource. It provides a significant amount of the World's energy supply and it is the basis for many industries. However, in areas where coal lies close to the Earth's surface and has been exploited by opencast techniques, radical alterations of landscape and significant impacts on the environment have occurred. This report was prepared to provide guidance to those who are responsible for the prevention of environmental effects from surface mining and for the restoration of the mining areas. Environmental problems of surface coal mining and restoration of the mine sites are discussed in the report. Particular attention is given to Eastern Europe, which continues to be a major centre of opencast lignite mining. Reclamation of mined lands for forestry, agriculture and wildlife is briefly discussed. However, the shear volume of coal removed from many mines in Eastern Europe is so vast, that there is often insufficient overburden material to refill the pits. Consequently, the main focus of this report is on the creation of lakes in these former surface mines. Many problems have to be overcome in creating healthy lakes for recreation or wildlife. Guidelines for treating water quality problems and further development of lakes are provided. Techniques for dealing with acidic waters, eutrophication and contamination are discussed. Examples from different countries are included in the report to help the reader to better understand different environmental problems associated with the surface mining and to appreciate the methods for restoration of the surface mine sites.

1
Environmental Problems caused by Surface Coal Mining

The former state of East Germany was the largest producer of lignite in the World, previously mining over 300 million tones per annum. Much of the lignite mining was centered in two regions: the central states of Saxony/Saxony-Anhalt and the eastern Lusatian region. In addition to tremendous destruction of the landscape, the mining caused massive alteration of surface and groundwater. It swallowed up entire rural landscapes complete with towns and villages and caused changes to the water table, e.g., causing the water table to drop by 30 m in places. Further environmental problems were caused by the industry supplied by the coal: sulphur dioxide and dust from power plants, emissions from chloride plants, organo-sulphuric and organo-nitrogenic compounds from pyrolysis. With lignite production now reduced to 100 million tons per annum many surface mines are closed. Many are being naturally flooded by the rising water table, but are acidic due to oxidation of pyrite. Others are filled with industrial wastes, often of poorly understood chemical speciation.

Massive lignite mines also exist in the northwestern region of the Czech Republic in the North Czech Brown Coal Basin and in the Basin of Sokolov. For example, near the town of Most in the North Czech Basin, there are six surface mines of total surface area 3,600 ha. These pits are up to 200 m deep, and since coal seams are up to 50 m thick in places, there is not sufficient overburden material to refill the mines. Environmental problems are similar to those in eastern Germany, again being increased by the high population density and developed industry in the mining region.

Poland is the third highest producer of lignite in Europe, mining around 65 million tons per annum. The coal is exploited in four large basins: Bełchatów, Konin and Adamów in Central Poland and Turoszów in the southwest – a region close to the German and Czech borders, known as the *Black Triangle of Europe* due to environmental degradation. The area of impacted terrain in Poland amounts to 60,000 ha. Much of the surface mining in Poland will continue to the middle of the 21st century. However, in western Poland there is already an artificial lake district of considerable size, containing over 100 man-made lakes. Several of these lakes are of acceptable water quality, having been formed by the collapsing of underground mines. Nevertheless, many of the lakes have established in former surface mines and are strongly acidic with high concentrations of iron and sulphates.

The Slovak Republic is relatively poor in coal deposits, but has a long-history of mineral mining, which has also given rise to environmental problems. All of the economically significant brown coal deposits are located in the Neogene Basin. From 1981 to 1988, exploitation of the open-pit mine Lehota in this region caused damage to agricultural land and 110 private houses. Examples of environmental problems relating to the mineral mining in Slovakia include: the collapse of a mining corridor between blocks of flats in a housing area located in the town Rudňany – Zápalenica; and damage to agricultural soil and forests by dust and gas emissions during magnesite mining and processing in the region of Jelšava.

The UK is an example of a country outside of the focus region with notable surface mining. The UK has a long history of coal mining, and though much of it is underground, the percentage of opencast mining has increased in recent years, having reached 30 % in 1985. Generally possessing smaller mines with larger overburden to coal ratios, the environmental problems in the UK are not as horrific as those in eastern Europe, but are nevertheless, of concern. Much environmental awareness is shown in current reclamation schemes, with attention, for example, to greenways and animal habitats of restored landscapes.

In North America, where extensive surface coal mining also occurs, the surface mining reclamation schemes of Alberta, Canada were of particular interest to this study. During the 1970s, a few surface mines in the Alberta foothills were adopted as recreational lakes, rather than being back-filled with soil at greater expense. Detailed study of several lakes found that the artificial lakes were able to support sport fish populations and that growth rates of the fish were superior to those in adjacent natural water bodies. New species of water-oriented birds were also attracted to the area of lake development.

2 Alternatives for Reclamation

For impacted lands surrounding mining lakes, or for regions where lake formation is difficult or disallowed, there are well-established alternative land-uses: forestry, agriculture, and recreational areas. The choice of reclamation scheme depends primarily upon climate (including microclimate), post-mining topography, existence of topsoil, proximity to urban centers and legal restrictions.

Physical soil properties, especially bulk density, infiltration capacity, rock fragment content and clay mineralogy are of importance to most reclamation alternatives. Soil compaction is often a key problem to overcome; it impedes plant growth and reduces infiltration capacity, potentially causing increased runoff and soil erosion. Consequently, the machinery used in recontouring is important; dump trucks cause less compaction than rubber tyred earth moving scrapers or dozers.

Chemical soil deficiencies can often be overcome with additives, e.g. fertilizers. However, long term chemical application is undesirable. A successful reclamation scheme should not require long-term intensive management to maintain a stable landscape.

2.1 Forestry

Forestry may be the predominant reclamation scheme for timber production, for aesthetic reasons or where slopes are too steep for agricultural use. If trees can be planted without re-grading the soil, then considerable cost savings can be achieved. Where soils are too compacted, then techniques such as deep ripping are required, but re-grading may then be necessary to create access for the decompacting machinery. Soil chemistry, and in particular pH, is important in species selection; knowledge of trees and shrubs that tolerate acid and alkaline conditions is well established. Hardy, soil-building nurse-trees should be included in the establishment of mixed forest, with additional consideration of soil moisture conditions. Choice of a single pine or conifer species is economically preferable for commercial plantations, but selection of three or more species reduces potential impacts of disease or insect damage. Herbaceous ground cover may provide supplementary nitrogen and stabilize toxic soils, but it will also compete with trees for essential nutrients, and so be detrimental to growth. Fertilizers, lime, mulches and fungal rhizom symbiontes (mycorrhiza) may be applied to improve growth during early years.

2.2 Agriculture

From an agricultural perspective, reclaimed lands may be used for pasture, cropland or rangeland. Productive pasture can be established by planting a single grass species or simple mix, with minimum effort and expense. In the seeding process there has to be a trade off between the classical requirement of a firm seedbed and the desire to increase infiltration potential and improve the rooting zone by tillage treatment. Chosen species should seed vigorously, be long-lived, adaptable to a range of conditions and be resistant to disease, insects and drought. Legumes are the primary plant species used on reclaiming cropland as they improve soil tilth, soil nitrogen and microbial activity. Other plant species include forage grasses, small-grain crops, row-crops, fruits and vegetables. Wheat adapts particularly well to mined soils due to its shallow root system. Reclaimed grange land has the potential to be more productive for livestock than pre-mining. Establishment of grasses and forbs to supply high quality feed and erosion protection is essential. In semi-arid regions, tillage tools, mulching and irrigation in early years may be required to supply and control water. Trees and shrubs are planted on reclaimed rangeland to enhance the utility of wildlife.

2.3 Wildlife

Other wildlife enhancement schemes used in any of the reclamation alternatives include planting of fruit bearing shrubs, critical placement of brush piles, construction of water resources and creation of boundaries. Choice of vegetation depends upon the habitat and food preferences of the target wildlife species.

3
Generation of Lakes in Former Surface Mines

There are two distinct alternatives for developing mining lakes in former surface mines. For human uses, such as drinking water, bathing or fisheries, the lake water should be as clean as possible. This is most easily achieved with deep, steep-sided lakes with low shore development, low nutrient inflow and phosphorus binding matter on the lake bottom. However, from the perspective of nature protection, lakes should have a high diversity of biological species. Such a target is achieved with a diverse morphometry. The lake should posses both deep and shallow parts, a long shoreline relative to lake area, bays, islands and mixtures of steep and shallow banks. High nutrient import and recirculation increase the bioproductivity.

3.1
Water Origin

In regions where the lignite and overburden have low sulphur content and sufficient carbonate hardness for acid neutralization, groundwater is best suited as the filling water. However, reliance on naturally rising groundwater can be disadvantageous due to the long filling time and potential problems of slope stability. Where available and suitable, pumping of groundwater from the dewatering of nearby active mine sites is preferable. Such a scheme is currently being used in the mining region south of Leipzig, Germany.

Filling of surface mines with surface water has the advantages of being relatively quick and usually involving neutral waters. The first of these advantages furthermore aids slope stability and causes the surrounding underground environment to become anoxic, which immobilizes metals in a sulfidic form.

The potential disadvantages of filling with surface waters are that they may be highly loaded with nutrients, oxygen consuming substances or hazardous substances. Subsequent problems are the development of algae due to eutrophication, reduced habitat for fish in deep water and contamination of groundwater below the pit, respectively.

To alleviate such problems a former surface mine near Bitterfeld, Germany is now used as a flow through water treatment plant for the River Mulde. The exiting water is more appropriate for lake filling water than the River Mulde upstream, since phosphorus is separated by sedimentation in this new reservoir. The main water quality problems to consider when planning the lake filling water are:

- acidification – low pH, e.g., due to pyrite oxidation, is mainly a problem for groundwater filling,

- salinization,
- contamination – e.g., where the mine has been used as a dumping site for industrial or urban waste,
- eutrophication,
- saprobization,
- infection with pathogens.

3.2 Living Conditions

There is usually little growth of bacteria and algae on mining lake beds given the existence of minerals with low organic carbon, but high phosphorus binding ability. Geogenic sulphur acidification is one of the most important factors effecting living conditions. Oxidation of pyrite has given rise to mining lakes with pH 2 – 3. Such lakes are only colonized by a few specialist algae and bacteria – those with mechanisms for detoxification. The pioneer plant *Juncus bulbosus* can appear soon after the first quick filling stages of an acid lake. The few animals found at low pH include rotifers.

3.3 Succession of Water Quality

The water quality of a mining lake may undergo considerable change during lake formation. For example, in the first few years the succession from a shallow to a deep water lake can cause oxygen depletion in the deep water. As a further case, when mining wells are first turned off, a fast rising of groundwater can bring acidic waters in to the lake. If surface water is then introduced the hydraulic gradient is reversed thereby avoiding acidification. However, when the lake is full, the gradient between the water table and the lake level may switch again, causing a second acidification.

When lake waters are contaminated with hazardous substances, each case is usually unique. The nature of the contaminant must first be established: inorganic vs. organic, degradable or not, solubility, possible introduction into nutrient chains. A key question then is what special conditions have to be established for functioning degradation and to allow the restoration of the contamination by in-situ self-purification.

3.4 Sediments

Investigation of sediment profiles can provide valuable information on the history of a lake development, such as annual sedimentation rates and changes between anoxic and oxic conditions. Analysis of the sediments in acid mine lakes has often shown a carbon content that is too small for normally functioning aquatic ecosystems. It has consequently been hypothesized that carbon is limiting the life starting processes in these cases, even when there are exposed coal faces at the lake bottom. The importance of iron-reducing bacteria in enhancing microbial alkalinity in sediments influenced by acid mine drainage has been confirmed with the help of sediment profiles taken from a research lake in the Lusatian mining district of Germany.

3.5 Costs/Benefits

There are many benefits that may offset the cost of developing mining pit lakes. Mining regions are typically areas of high population with excellent infrastructure, but often high unemployment. Against this background lakes provide possible nuclei, attracting water-oriented tourism, including fisheries. To evaluate the economic suitability of alternative lake developments, the following decision criteria may be used:

a) natural fit of the landscape;
b) development for recreational purposes; and
c) restrictions due to environmental and other impacts.

Recreational benefits can be expressed in a monetary form by accounting for the number of recreational days (per year) created and the costs of one recreational day in an artificially built open-air basin (e.g. outdoor swimming pool). A possible direct source of income at mining lakes comes from the sale of sport fishing licenses.

4 Controlling and Improving Water Quallty

4.1 Acid Mine Lakes

The creation of an acid lake can be avoided if actions are taken from the initial stages of mining operations. Dewatering of the mined lignite can be controlled to minimize the exposure time of pyritic minerals with oxygen. Acid causing overburden heaps should be placed downstream of the future lake relative to the greater groundwater field. The initial filling of the lake is important, and should be conducted so as to reduce hydraulic gradients of acidic waters into the lake. Revegetation of acid heaps can be boosted by treatment with alkaline material such as lime or ash, thereby reducing infiltration. Detergents, which kill acid forming bacteria, may be applied to smaller heaps rich in sulfidic ores.

A number of in-lake approaches can be taken for neutralizing existing acid lakes. Lake stratification is a desirable condition for acid reduction; it can be induced by placement of oil barriers, floating reeds or submerged plastic sheets into the open water. Surface waters containing neutralizing and buffering bicarbonate, nutrients and organic matter can be added, but not in excessive amounts. The addition of phosphorus may also be required to promote natural neutralization processes. A gentle fertilization may be useful for the restoration of the food web in acidified lakes. Liming has been successfully applied in several countries to neutralize rain-acidified lakes. The addition of organic matter, such as potato chips or waste from a sugar factory may enhance the sulfide forming properties of the sediments. A further anaerobic ecotechnology is the placement of a bale of straw, pumped through with nitrate rich surface water and fatty acids as the food source.

Ex-situ treatment of acidic waters include fully or partially anaerobic systems and anoxic biofilm chambers. Constructed wetlands or oxidation ponds may serve as the aerobic polishing of anaerobic neutralized waters.

4.2 Eutrophication

Control of eutrophication is high priority in neutral mining lakes when good water quality is sought. However, for shallow mining lakes in protected natural landscapes, where the aim is high species diversity, eutrophication is not a drawback. To protect against eutrophication, mining lakes can be designed with great depths and steep banks, etc., although there has to be some trade-off with slope stability.

Once a lake has already formed, control of eutrophication is mainly achieved by reducing the import of nutrients, and in particular phosphorus. This may be through sewage treatment plants, storage reservoirs for nutrient elimination, or planting of forest strips to reduce nutrient export from agricultural land. Phosphorus elimination facilities at lake inlets are available, but due to the short operation for filling too expensive and currently at the research and development stage.

A number of in-lake measures against eutrophication may be considered. Top-down biomanipulation of the nutrient pyramid can lead, for example, to a reduction of phytoplankton in a lake. Precipitation of nutrients, e.g., using aluminum sulphate, is a further alternative, although it is expensive. Desludging is a further option, but it can be technically difficult. Further methods include deep-water aeration, delivery of deep water and harvesting of macrophytes. Application of copper sulphate should only be considered in extreme situations.

4.3 Contamination

Opencast mining holes have been used for the deposition of a vast range of industrial and municipal wastes. Each deposit is usually unique, requiring detailed chemical speciation. Besides the chemical analysis, ecotoxicological effects should be investigated with help of an inventory of indicator organisms. Liquid deposits are generally more critical than solid materials, as they may potentially spread into the groundwater. The most important question for the assessment of impacts is whether or not degradation takes place at all.

A scheme has been developed outlining different approaches for treating contaminants in and around mining lakes in eastern Germany, using ecologically acceptable and economically feasible methods. Contamination may be removed, isolated or restored, using in-situ or ex-situ treatments, employing physical, chemical or biological methods. Examples of problems tackled include the removal of a waste dump at Nachterstedt, isolation of a landfill at Halle-Kanena, and subhydric incapsulation of a complex film factory sludge at Wolfen. Most ex-situ treatments are expensive and generally not satisfactory.

A number of in-situ ecotechnologies have been demonstrated as pilot-scale experiments and beyond. These approaches aim only to help the system help itself by metabolism and cometabolism, incorporation, bioaccumulation, bioflocculation and sedimentation of the treated materials. After conducting pilot scale tests at the Schwelvollert Lake in Germany, phenol, ammonium and humus-like contaminants were finally treated in the whole water body by flocculation, liming and addition of phosphorus. This was one of the first full-scale examples of the bioattenuation/bioremediation of an industrial wastewater deposit.

Contents

Introduction

Ulrich Stottmeister[1], Alena Mudroch[2]

[1]UFZ Centre for Environmental Research Leipzig-Halle, Department of Remediation Research, Permoserstr. 15, D-04318 Leipzig, Germany
[2]National Water Research Institute, 867 Lakeshore Rd., P.O. Box 5050, Burlington, Ontario L7R 4A6, Canada

It is generally accepted that significant degradation of the landscape occurs during the mining of most raw materials. The extent of the impact depends on the type and scale of the mining operation. Mining of coal and lignite, two of the most important materials for generating energy, has affected large areas in many countries. Past coal and lignite mining carried out on a small scale had little impact on the environment. Nature recovered from the negative influence of such small-scale mining activities without any human help. However, in the 19th and 20th centuries gigantic open pit coal and lignite mining was introduced in many countries to generate sufficient energy for an increasing world population. The technology of the large-scale open pit coal and lignite mining has been accompanied by radical alterations of landscape and significant impact on the environment.

Many different transformations occur in the surface appearance, geochemical environment, atmosphere, surface and ground water, soil and some components of the physical environment during removal of overburden, mining the coal seam, construction of waste dumps and treatment of the coal. Impacts on biotic components of the ecosystem involve not only flora and fauna, but also mankind.

The effects of many years of open pit coal and lignite mining are well documented for the area called “The Black Triangle of Europe” located near the borders of Poland, Germany and the Czech Republic. Following are selected examples of the effects of open pit coal and lignite mining in this area.

Coal mining in the Upper Silesian Coal Basin in Poland started approximately 200 years ago. By the end of the 1970's, the level of production of hard coal in this region reached 200 tons per year. Many coal seams have been exploited over a large area of the Basin. This has resulted in subsidence in the area averaging several meters and in some regions has reached 30 m. In addition to the deformation of the landscape, the coal industry has also had a detrimental impact on groundwater. It was estimated that the total volume of water draining Upper Carboniferous formations is approximately 100 km^3. From this water pumped from the coal basin, approximately 670 m^3/min is used for coal processing while a further 450 m^3/min is discharged to surface water. The total salt discharge (i.e., dissolved chloride and sulphate ions) is estimated to be approximately 7,400 tons per day. In addition, there are over 140 waste dumps, which are subject to leaching and wind erosion. Mining in the Basin has an effect on surface waters, shallow subsoil waters and the deep water-bearing horizon in the vicinity of the workings. All mines in the coal basin discharge sewage, industrial waste and mine water to surface water.

The first small coal mines operated in the northern Czech Republic in the 15th century. In the 20th century, significant increases in coal mining commenced after a recommendation by the government to replace expensive and fast diminishing wood supply by another heating source. During the Second World War, new technology was implemented for open pit coal mining. The coal was used to produce synthetic gasoline necessary for the war activities. The increase in coal mining activities in the Czech Republic continued after the war to support the high-energy demanding economy, particularly to produce electric power, steam and heat. This large increase in open pit coal mining brought about a considerable and irreversible negative impact on the ecosystem in the mining areas.

The coal mining industry has a long tradition in eastern Germany. For example, records show relatively intensive coal mining in the late 17th century in the area to the south of Leipzig. Brown coal was the basis for the development of local industry. More than 20 open pit mines existed in 1863 in the region around Borna. The open pit coal mine Espenhain started in 1937. The total amount of mined coal at this mine was approximately 568 million tons. Approximately 1,700 billion tons of overburden was moved at the area. The average depth of the large pit was 90 m. During the mining, approximately 3,980 ha of land were devastated. Mining had an additional significant negative impact on the environment due to the use of many open pit mines in the Halle-Leipzig region for deposing of different industrial wastes.

Due to recently improved mining techniques and new conceptions for preserving the environment, the impact of open pit coal mining on landscape and the environment is decreasing. Several countries have already implemented legislation to prevent or minimize the negative environmental impacts of open pit mining. As a result of efforts to rehabilitate areas where open pit mining has terminated, farmers and foresters now successfully use thousands of hectares of re-cultivated land. Many abandoned open pits have been or are being changed into lakes. Additional design the landscaping of areas adjacent to the generated lakes improve the remediation of the coal mining sites into areas which can be used for recreational activities and as wildlife, particularly waterfowl, habitat.

The objectives of this project were to contribute to the prevention of environmental impacts of surface coal and lignite mining, and to promote innovative technology for restoration and remediation of the mining sites, in particular the generation of lakes in former open pit mines. This is done to support sustained economic and environmental benefits between NATO members and other countries. The description of restoration/rehabilitation of the mining areas and design for development and proper management of lakes generated from the mining pits given in this report are complemented by case studies for different countries. Thus, the report provides guidelines for the countries which need to develop strategies for the rehabilitation of open pit coal mining areas prior to the commencement of the mining activities, as well as for the countries experiencing environmental problems at open pit coal mining sites. Many recommendations and observations given in the report are additionally applicable to areas of surface metal mining.

The following topics are discussed in this report[1]:

- Surface coal mining problems and some examples of their solution in selected countries, Germany, the Czech Republic, Slovak Republic, Poland, U.K. and Canada;
- Restoration/rehabilitation options for mining areas, particularly restoration for forestry, agriculture and wildlife with a detailed description of methods for generating lakes at past surface coal mining areas;
- Problems of acid mine drainage associated with the generation of the mining lakes, and the restoration of such lakes;
- Experience with solving problems of eutrophication and contamination of mining lakes;
- Development aspects for the post-mining landscapes in Central Germany
- Social aspects of living next to a opencast mine, and
- Art projects in post mining landscapes, as a memorial and signal for new beginning.

[1] The Chapters *Executive Summary* (Kennedy), *Introduction* (Stottmeister, Mudroch), *Reclamation and Regeneration of Landscapes after Brown Coal Opencast Mining in Six Different Countries* (Stottmeister, Mudroch, Kennedy, Matiova, Sanecki, Svoboda), and *Alternatives for the Reclamation of Surface Mined Lands, Mining Lakes: Generation, Loading and Water Quality Control* (Klapper) were published as UFZ report 11/1999, ISSN 0948-9452.

Reclamation and Regeneration of Landscapes after Brown Coal Opencast Mining in Six Different Countries

Ulrich Stottmeister[1], Alena Mudroch[2], Christopher Kennedy[3], Zdena Matiova[4], Jacek Sanecki[5], Ivan Svoboda[6]

[1]UFZ Centre for Environmental Research Leipzig-Halle, Department of Remediation Research, Permoserstr. 15, D-04318 Leipzig, Germany
[2]National Water Research Institute, 867 Lakeshore Rd., P.O. Box 5050, Burlington, Ontario, Canada L7R 4A6
[3]University of Toronto, Department of Civil Engineering, 35 St. George Street, Toronto, Ontario, Canada M5S 1A4
[4]HGM, Ltd., Velka Okruzna 38, 010 01 Zilina, Slovakian Republic
[5]Polish Academy of Sciences, K. Starmach Institute, Freshwater Biology, ul. Stawkowska 17, 31-016 Kraków, Poland
[6]R-Princip Most, s.r.o., Budovatelu 2830, 434 01 Most, Czech Republic

1 Problems Resulting from Lignite Mining in Eastern Germany

The states of Saxony and Saxony-Anhalt in central Germany have a long history of heavy industry. This tradition was continued following the foundation of communist East Germany in 1949. However, throughout its lifetime, much of East German industry remained of pre-war vintage. Until the regime collapsed in 1989, efforts were primarily directed towards boosting industrial output, with scant attention being paid to the environmental costs. A similarly negligent attitude prevailed in Poland, Czechoslovakia, and the other Eastern Bloc countries.

East German industry greatly relied on lignite (brown coal). It was used to fire the country's power stations, producing the electricity needed, for example, by the chlorine-based chemical industry and the country's steelworks. Lignite was also decomposed by heating it under the exclusion of air (a process known as pyrolysis) to produce raw materials for the chemical industry.

After World War II the newly founded East German State (the GDR) built upon its industrial legacy by intensifying mining and industrial activities. Until its economic collapse in 1989, the GDR was the world's largest lignite producer, mining over 300 million tons from 33 open pits per annum.

Lignite was mined in the areas surrounding the industrial centres (Fig. 1).

In the east, the Lusatian mining district (about 60 x 60 km) was dominated by opencast mines, while in the western GDR, the central German lignite-mining district, in particular the area around the cities of Leipzig and Halle, measured about 90 x 40 km.

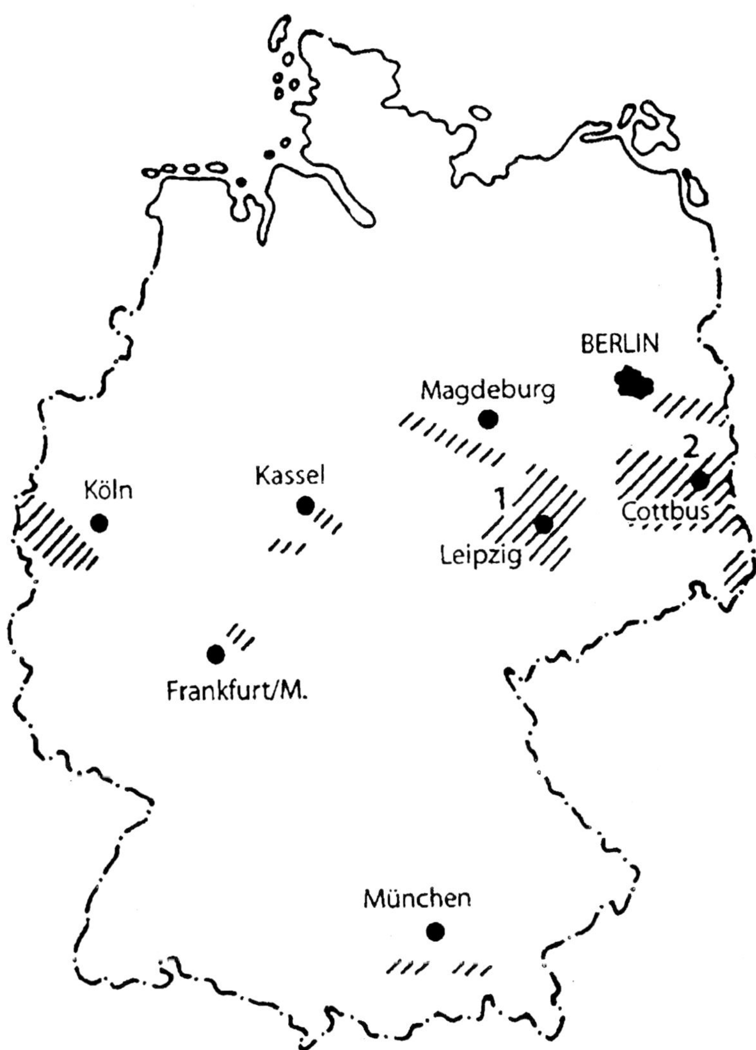

Fig. 1. Lignite mining areas in Germany (1: Saxony, 2: Lusatia; Stottmeister et al. 1999)

Lignite mining resulted not only in a tremendous destruction of the landscape, but also in the massive contamination of surface water and groundwater. The mining had an enormous ecological and sociological impact. It swallowed up entire rural landscapes complete with towns and villages and caused the water table to drop by more than 30 m in places, thereby affecting the water balance and destroying the landscape with little prospects of adequate restoration in sight.

Moreover, emissions from the now obsolete industrial plants impaired the population's health and quality of life. Sulphur dioxide and dust from power plants and domestic stoves, emissions from chlorinating plants, and organo-sulphuric and organo-nitrogenic compounds from pyrolysis were all widespread. Meanwhile the rivers served as sewage pipes for untreated industrial effluent.

However, lignite mining and processing was not solely to blame for the harm suffered by the environment. Other culprits included uranium mining, the potash mines and copper extraction.

The upshot is that Germany now faces a number of colossal challenges – namely the recultivation of ruined landscapes, and the decontamination of the soil, groundwater, rivers and lakes. One especially formidable problem is posed by the clean-up of sites polluted by organochlorines and hydrocarbons.

After German reunification, lignite-mining was drastically reduced and lignite-processing abandoned, causing the environmental situation to rapidly improve. A restoration programme was set up by the Federal German Government to assess the impacts of the former lignite mining. Within the space of just five years (1989 – 1994), annual lignite extraction totals in central Germany and Lusatia have dropped from about 330 million tons to some 100 million tons. The current output target in these two districts is around 70–90 million tons per year.

The following three scenarios typify the current state of abandoned opencast lignite mines in eastern Germany. (No consideration is given here to the lignite mining in the western Rhine River region of Germany, which has further conditions and problems.)

- Open pits are being or have been flooded to form recreational lakes, resulting in a new type of landscape. The new lakes generally do not influence the groundwater levels, but probably the microclimate. Problems include mechanical slope erosion in the shore zones and the acidification of the water by natural oxidation processes (forming "acidic lakes").
- Opencast mines are being naturally flooded, normally by the rising water table and by rainfalls. However, the water is contaminated by uncontrolled municipal waste deposits adjacent to the lake or by agrochemicals and fertilizers.
- Disused opencast mines were used for waste disposal, especially

industrial waste without any safety measures. The nature of the material dumped is often unknown and so it must first be analysed. Studies have to be carried out into the chemical interactions and the formation of metabolids so that proposals for remediation and protection can be drawn up.

Many specific examples of mining lake problems in eastern Germany are discussed later in Sections "Mining Lakes:..." by Helmut Klapper.

2
Reclaiming the Great Surface Coal Mines of the Czech Republic

Restructuring of the Czech coal industry is having a positive impact on the environment in opencast mining areas. This has mainly resulted from the establishment of new environmental legislation and incentive economic mechanism improving and preserving the environment. Adequate planning and implementation of ecologically-sound practices is considered a major prerequisite for improving the environment and quality of life in the Czech opencast mining areas.

The Czech Republic has substantial reserves of high-energy brown coal. This natural resource, exploited almost exclusively by opencast technology, occurs in two separate basins in the northwest part of the Czech Republic. The larger basin is called the North Czech Brown Coal Basin, and the smaller is the Basin of Sokolov (Svoboda 1993).

The conditions and problems of mining in both basins are very similar and relate mainly to the complex geological structure. The basins are characterized by the unevenness of coal seams that are pervaded by many irregularities. The areas of the basins are deeply disturbed by past underground mining operations. In addition there is a diverse composition of overlying soils.

Furthermore, the coal is mined in an area with high population density, developed industry, a dense transportation network and other different structures. Therefore, the mining enterprise is continuously required to solve problems regarding clashes of the mining activity with other interests in the region. With a high demand for coal, surface mining became the dominant technique used in coal exploitation.

The strategy in the past was to concentrate the surface mining into a small number of super-mines with modern continuous production technology. Prior to 1989, such intensive mining activities were the result of a national economy with a strategic goal to develop a strong heavy industry and other sectors requiring high energy input.

A significant decrease in mining occurred in 1989 under the new market conditions following the political/economic changes. Nevertheless, as a legacy of the socialist past, the Czech Republic was left with highly damaged territory on which the concentration of industrial production had exceeded the ecological capacity of the region. Many historical buildings, villages and cities were torn down to generate the large super-mines. Occasionally, buildings of special note were saved. Fig. 2b shows a historical cathedral that was transported over several hundred meters to make space for the expansion of one of the super-mines. The opencast mines generated in the Most area of the northern Czech Republic, where the cathedral is situated, are also shown in Fig. 2.a-d. The result of the mining is complete devastation of the landscape.

However, at present, the mining activity is organized in such way that the exploited land will be restored at the earliest possible opportunity. This time period for full restoration is estimated to be approximately 20 to 30 years, but, longer in some areas.

One of the most important tasks for the ecology, and also the economy, of the Czech Republic is a gradual reduction of the area of the land devastated by the coal mining. This is being achieved by a reduction in the number of mining sites and by the implementation of fast and effective methods of land restoration in areas that are no longer used for coal mining or spoil disposal. Presently, the effects of opencast mining on the landscape are being addressed by different methods of re-cultivation, such as agriculture, forestry, recreation and amenities. Fig. 3a shows a mining area restored into a recreational lake in the Most area of the northern Czech Republic. A racecourse was built in the same area after filling a large surface mining open pit with overburden material (Fig. 3b).

There are six large pits of total surface area of 3,600 ha in the North Czech Basin. The maximum depth of these pits is 200 m. Three further pits in the western basin have a total area of 2,000 ha. The coal seams in parts of these mines are up to 50 m thick. Consequently there is not sufficient overburden material to refill the pits.

For economic reasons, it is proposed to rehabilitate these pits by flooding with water. The greatest technical problem to be overcome in the proposed plan is establishing the quality and quantity of the flooding water. The North Czech Republic coal basin is drained by the river Bilina with an average discharge of 4 m^3/sec.

Furthermore, the river has poor water quality, particularly at low flow rates. The situation in the West Czech coal basin is better. There the River Ohre has a flow of 4 to 5 times that of the Bilina and its water is of a better quality. Pumping of the water from the River Ohre to the North Czech mines is an option under consideration.

Fig. 2a.

Fig. 2b.

Fig. 2a-b. Surface mining in the Most region of the Czech Republic (b: transported cathedral)

Fig. 2c.

Fig. 2d.

Fig. 2a-d. Surface mining in the Most region of the Czech Republic

Fig. 3a. Recreation area besides mining pit lake

Fig. 3b. Race course

Fig. 3a-b. Reclamation schemes at Most, Czech Republic

3
Environmental Impacts of Surface Coal and Mineral Mining in the Slovak Republic

The Slovak Republic is relatively poor in coal deposits in comparison with Germany, the Czech Republic and Poland. All of the economically significant brown coal or lignite deposits are located in the Neogene Basin. The only surface mine of economical importance in this basin is the Lehota mine in the Novaky deposits. In addition to this mine, there are a few smaller surface coalmines of local significance, from which the coal was exploited for the local use in the past.

Coal mining activities in the Handlova-Novaky coal basin are unfortunately affected by water inflow from water-bearing systems in underlying beds. Water breakouts occurred in the past.

Arsenic emission from coal incineration at the Novaky power station is of concern. Conclusions of a mineralogical study showed that the main sources of arsenic in the coal are the minerals realgar and auripigment. From a hydrogeochemical point of view, arsenic is commonly observed in the region of Upper Nitra in the Slovak Republic, particularly in the “old water” where its concentration reaches 1.2 mg/L. Fluorine compounds are also found in high concentrations.

The exploitation of open-pit mine Lehota from 1981 to 1988 had a devastating impact on the environment (Fig. 4). Much of the damage was to agricultural land. Slip faults of various sizes are common in the area as well as pits and terrain depressions, both water bearing and dry. The physical and chemical properties of the soil have been changed after degradation by water, flooding and different chemical reactions. In addition to degradation of the agricultural land, water regime changes have been observed. Also, there was a considerable impact on the local Lehota community including damage to 110 private homes.

Remediation of the Lehota mining area has so far involved recultivation. This began in 1992 with the refilling of pits in the south part of the mine. In 1994, there were 399 ha of agricultural soil recultivated at a cost of 153,090 Slovak korunas. In the future, there is the possibility of lake development in the Novaky deposit, as well as in the abandoned local coal mines of Backov, Liesek, Drienovec and Obyce. In addition to the above environmental problems generated by surface coal mining in the Slovak Republic, surface mineral mining has caused damage to the environment. Several of the environmental effects of the surface mineral mining experienced in the Slovak Republic are similar to those originating from the surface coal mining. Therefore, these effects are discussed below.

Mineral mining results in devastation of the environment from the following:

Fig. 4. Lehota pod Vtačnikom brown coal deposit

Fig. 5. Banská Štiavana "Terézia" tongue

Fig. 6. Rudňany: caved in "Droždiak" tongue in the locality of Baniská

- landscape morphology changes (quarries, settlement, mine dumps),
- changes of hydrogeological conditions (water table decreasing, inundation),
- changes of geochemical conditions (groundwater and surface water contamination and contamination of surrounding soils),
- accumulation of solid wastes (heaps, sludges),
- influence of explosive works (seismic effects, pressure waves, noise etc.).

The Slovak Republic has a mining history lasting several centuries. In the past many types of metal mining were undertaken. However, recently mining has been rapidly declining. Only two metallic mineral deposits are now in operation: a siderite deposit Nižná Slaná in the "Spišsko-gemerské rudohorie" mountains situated in the east part of Slovakia; and a gold deposit in Hodruša in the central part of Slovakia. Nevertheless, over recent years ecological problems relating to metal mining activities have occurred. Two serious ecological accidents occurred in the districts of Šobovo and Smolník (Šucha et al. 1996). When the metal mining industry first developed, only surface deposits were mined for the following reasons:

- metal mineral outcrops were easy to access and they could be mined without expensive mining technology,
- soft, oxidized hematite, an earthy ore that is easy to treat, could be easily exploited at siderite deposits,
- the oxidized zones were often enriched with gold and silver and were, therefore, of more economic interest.

Fig. 7. Rudňany: caved in "Droždiak" tongue in the locality of Baniská

Fig. 8. Šobov: the countryside damaged by acid waters below the quartzite deposit

The metallic mineral deposits were usually just mined down to the water table. Example cases are: the "Boží dar" tongue near Gelnica; the "Hrubá" tongue at the Slovinky deposit; the "Terézia" tongue in Banská Štiavnica (Fig. 5); and the "Droždiak" tongue in Rudňany (Fig. 6 and 7).
Evidence of surface mining can be observed at almost every metal mineral outcrop in the Slovak Republic. Abandoned mining pits of smaller dimensions covered by vegetation occur in several areas. Negative effects of mining can be observed at the well-known Rudňany deposit in the "Spišsko-gemerská" area. There, the important "Droždiak" tongue is located in the northern part of the deposit connected with a prominent tectonic slip line in the West Carpathians (length 6 km, average thickness 6 m, thickness at surface up to 30 m, depth down to 900 m in places). Barite dominates in the tongue down to depths of 200 to 300 m. The barite content decreases with depth, whereas the contents of siderite and quartz increase with the depth. In the past siderite-sulphide ores have been mined there. Later, barite was mined at the surface as can be seen at Baniská (Fig. 6 and 7). Today, there is large abandoned quarry in the past mining area. Many faults and large earth movements have also resulted. For example, there is a mining corridor between blocks of flats in a housing area located in the town Rudňany-Zápalenica, with mining activity down to depths of 80 m. During vertical tongue mining by the means of an upper chimney at a depth of 25 m, the chimney caved in. The pit that resulted endangered the housing area.

Tips containing spoil material and waste created from ore treatment, often concentrated in sludges, are the most frequently occurring environmental problems associated with the mineral mining.

A study by Šucha et al. (1996) has shown the risks associated with the pit tips and sludges in the areas with higher concentrations of unstable sulphides. Rapid decreases in soil pH values and accumulation of iron and aluminum oxihydroxides and sulphates takes place as a result of the unstable sulphides (Fig. 8 and 9). These effects result in serious breakdown of the soil function and its properties, in cases, leading to complete soil degradation and erosion. The underlying cause of this problem is the oxidation of the sulphides present in pit tips. The oxidation rate mainly depends on the type of the sulphide material. Pyrite and pyrrhotine can be broken down most rapidly. Products of the oxidation transferred to the environment are often toxic. Further information on the impacts of mineral mining on the environment at the localities of Pezinok, Smolník, Banská Štiavnica-Šobov, Rudňany and Slovinky is presented by Šucha et al. (1996).

Fig. 9. Šobov: the countryside damaged by acid waters (foreground) from a quartzite deposit (background)

The mining and processing of magnesite has further negative influences on the biosphere, particularly on the soil and vegetation. The magnesite industry in the Slovak Republic includes units in Jelšava, L'ubeník, Hazava and Hnúšt'a. Agricultural soil and forests are damaged by dust and gas emissions in these regions during magnesite mining and processing. Harmful manganese components are released into the atmosphere as solid emissions, which can effect the microclimate of the nearby region that have an effect as deposits under specific conditions. Included in the mixture of Mg-emissions is the very reactive manganese oxide arising at temperatures in the range of 800 to 1100 °C. Surface and underground magnesite mining has further negative effects, such as subsidence, landslides, microseismicity, erosion, etc., causing changes in the hydraulic regimes of surface and groundwaters, as well as their quality. The mining activities in the regions described above have resulted in some spectacular changes to the rock environment. The result of a collapse at a magnesite mine near Jelšava is shown in Fig. 10 and 11.

Fig. 10. Magnesite deposit at Jelšava

Fig. 11. Collapse at the Jelšava magnesite deposite

4
The Mining Lakes of Western Poland

Poland is one of the main producers of lignite in the world. Annual exploitation reaches 63 845 000 tons, constituting approximately 10 % of the European production or 6.9 % of the world production. This places Poland as the third highest producer of this raw material in Europe and fourth in the world (Statistical yearbook 1997). Lignite, along with pit-coal, provides the basis for Polish industry. Consumption for energy supply amounts to 62 887 000 tons.

4.1
Current Areas of Exploitation

Mining of brown coal in Poland developed on a large scale in the years after the Second World War and is carried out by means of opencast techniques. This leads to damage of the landscape and biocoenosis of remarkable areas. Exploitation of brown coal causes a complete devastation of soils. Hydrological relationships are subject to transformation, both within the exploited terrain as well as in the surrounding catchment area.

Currently, brown coal is exploited in four large coal basins: Betchatów, Konin and Adamów in Central Poland and Turoszów in southwestern Poland near the confluence of borders with Germany and the Czech Republic. Due to the environmental degradation, the region close to the borders of these three countries is known as the "Black Triangle of Europe."

It is predicted that the brown coal exploitation in Poland will continue until the middle of next century. However, currently recultivation of the destroyed terrain is conducted on a wide scale. This mainly involves forestation of arising waste-tips and partial filling of opencast pits in the course of further exploitation. However, the problem of recultivation of these environments will not take place in its final form until exploitation of the currently active mines is finished. It is therefore necessary for Poland to follow intently the world experiences in this field to help formulate its reclamation plans for when the mining is finished. Such plans will be of a large dimension for the area of degraded terrain in Poland amounts to 60 thousands ha (Guziel 1988).

4.2
Areas of Abandoned Exploitation

Within today's Polish borders there are some areas where brown coal was exploited in the 19th century. These are generally in the western region of the country, which was part of Germany prior to the Second World War II. The main region is situated in the province of Zielona Góra under the name of Łuk

Mużakowski. Lignite was exploited there by means of underground methods in addition to opencast mining. After the underground mining, there was no back filling of the mine. Only wooden supports were left, many of which have now collapsed. Consequently, the soil surface has settled forming long kneading-troughs or funnels, which have filled with water. Natural forest stands and arable land in these places have been destroyed (Jędrczak 1992). In this post-mining area there are now a little over 100 artificial lakes of total area over 150 ha. The lakes range from about 25 to over 100 years in age and the area has become known as the "Anthropogenic Lake District" (Kozacki 1976). There is further degraded terrain on the other side of the border with Germany, where 330 artificial reservoirs of total area 535.3 ha are situated (Pietsch 1970).

Polish scientists have investigated the characteristics of this water environment, in respect to both hydrochemistry (Matejczuk 1986; Jędrczak 1992) and hydrobiology (Matejczuk 1989; Koprowska 1995).

4.3 Hydrochemical Characteristics

The chemical properties of the waters in the many of the reservoirs differ from those of natural waters in the catchment basin. The concentrations of iron, and sulphates are increased and the waters are strongly acidic. The cause of this strong acidification is pyrite (FeS_2) associated with the lignite deposits.

Pyrite is subject to chemical and biological processes of decay in three stages (Baker and Wildshire 1970):

I. chemical oxidation

$$2FeS_2 + 7O_2 + 2H_2O = 2Fe^{+2} + 4SO^{-2}_4 + 4H^+$$

or biological oxidation

$$2FeS_2 + 7.5\ O_2 + H_2O = 2Fe^{+3} + 4SO^{-2}_4 + 2H^+$$

II. cyclical degradation

$$4Fe^{+2} + O_2 + 4H^+ = 4Fe^{+3} + 2H_2O$$
$$FeS_2 + 14Fe^{+3} + 8H_2O = 15Fe^{+2} + 2SO^{-2}_4 + 16H^+$$

III. precipitation of iron III compounds

$$Fe^{+3} + 3H_2O = Fe(OH)_3 + 3H^+$$

The speed of the decay process depends upon the pH. At pH > 4.5 this process takes place more quickly by chemical means. When the pH decreases below 4.5, oxidation runs more quickly by microbiological processes. At a pH between 3.5 and 4.5 oxidation takes place with participation of bacterium from genus *Metalogenium*, and at pH below 3.5 with participation of bacterium *Ferrobacillus ferrooxidans* (*Thiobacillus ferrooxidans*).

The most complete hydrochemical investigations in the area of the Polish "Anthropogenic Lakeland" were conducted by Jędrczak (1992). This physico-chemical research considered 63 of the 100 or so existing lakes. Two groups of lakes were distinguished:

- Acidotrophic with pH from 2.6 to 3.7 and redox potential from 605 to 755 mV.
- Near neutral lakes with pH from 5.2 to 7.4 and redox potential from 380 to 600 mV.

The low pH lakes formed where brown coal had been exploited by opencast methods and those with higher pH were created in depressions of soil where underground mines had collapsed. In the latter case the soil surface was not removed in the course of coal extraction.

In the group of acidotrophic reservoirs, of which 30 were examined, concentrations of sulphates ranged from 101 to 1260 mg/l. Content of iron was from 0.3 – 182 mg/l. These waters were poor in organic matters, their oxygen demand was in the range of 1.2 – 21.6 mg O_2 per litre. Concentrations of nitrogen amounted to 0.14 – 11.6 N-NH_4 mg/l and 0.05 – 1.81 N-NO_3 mg/l.

Most of the remaining artificial lakes had waters that were of class I or II purity.

The small river Chwaliszówka, which drains the lake district region had strongly acid waters of pH 3.5 – 4.2. The content of organic matter is low (COD 6.5 – 27.2 mg O_2 per litre). The quantity of sulphates is 108 – 155 mg/l, while the concentration of iron is very high (8.3 – 32.0 mg/l); such values are rarely met in freshwaters (Jędrczak 1992).

4.4 Hydrobiological Characterization

The physico-chemical water parameters of the acidotrophic reservoirs mentioned above do not allow the possibility for good development of biocoenoses. Only extreme biotopes and a small number of organisms are able to colonize such waters. This is further evident from the small number of algae taxa and invertebrates that were found.

A list of algae taxa identified in four acidotrophic lakes, in different phases of succession, and where pH was 2.8 – 3.2, contains only 19 taxons, of which 13 were diatoms (Koprowska 1995). In all four lakes the diatom *Eunotia exigua* was predominate. Most of the species of this genus are recognised as acidobionts. Other species were found in smaller numbers. There were diatoms – *Melosira granulata*, *Nitzschia palea* and *Cyclotella meneghiniana*, green algea – *Ulothrix* sp, cyanobacteria – *Lyngbya ochracea* and euglenophyte – *Euglena mutabilis*. The remaining species were adominants (Koprowska 1995).

The list of animals is yet poorer. Only one species, *Brachionus* sp., was found in the seston. In the benthos were two species of Chironomidae – *Chironomus plumosus* and *Limnophyes pusillus* and one species of Simulidae – *Sialis lutaria* (Koprowska 1995).

Not all species found were represented in each lake. In the lake of the most primary state of succession, the number of algal species amounted only to five and of the animals only *Chironomus plumosus* was found. In two lakes of more advanced succession 15 to 17 species of alga and 3 of larvae of insects were found. The larvae *Chironomus plumosus* did not attain large sizes (maximum 14 mm; Koprowska 1995). This confirmed the investigations by Ryhänen (1961).

Overall, it is important for Poland to address the problems of its relatively small acid lakes in the west, for when the far greater current lignite mines are closed down, they will constitute a considerable environmental problem.

5 Development of Lakes in Alberta, Canada

This contribution is a summary of an extensive report "Development of Sport Fisheries in Lakes Created by Coal Mining Operations in the Eastern Slopes" prepared in 1994 by Luscar, Ltd., Luscar-Sterco (1973), Ltd., Cardinal River Coals, Ltd., Pisces Environmental Consulting Services, Ltd., and Bighorn Environmental Design, Ltd., Edmonton, Alberta, Canada. We would like to express our thanks to the staff members of Luscar, Ltd., Edmonton, Alberta, for responding to our request and releasing the report to the members of our project team.

The Land Surface Conservation and Reclamation Act (LSCRA) was promulgated in Canada in 1973. Prior to 1973, the major goal of reclamation of surface coal mining areas in Alberta, Canada, was to stabilize the land surface and establish a vegetative cover. The reclamation did not consider returning the mining areas to their previous land uses or establishing specific land use goals. However, following the 1973 LSCRA, the primary objective for the reclamation has been the return of the mining areas to their pre-mining values and capabilities. The goal of all companies participating on the reclamation of surface coal mining areas in Alberta has been the creation of post-mining landscape suitable for multiple use. Creation of the lakes for recreational activities and potential fisheries has been added to regular reclamation activities that focus on returning productive forestry or wildlife land uses. These reclamation activities have also been supported by the following considerations. For mines operating at high topographic locations, such as those in Alberta, the cost involved in re-handling overburden for back-filling of mined out pits is often prohibitive. Therefore the creation of pit lakes may be the only economically feasible reclamation alternative. Lakes are preferred by the public because of their recreational potential for fishing, boating, swimming, etc., as well as for their aesthetic value. In addition, generated lakes are beneficial to the overall ecology of the rehabilitated mining area by providing wetland habitat for waterfowl and wildlife.

In the past, several ponds have been created by coal mining operations in Alberta. However, only few records exist about most of these water bodies. For example, in 1977, a mine pit near Forestburg, Alberta, was converted to a 7 m deep pond with a surface area of 6.2 ha. The pond was first stocked with rainbow trout in the spring of 1980. Data obtained from 1979 to 1981 did not identify any trends in water quality of the pond. Although the content of dissolved solids was high (in the range of 2,000 mg/l), it did not affect trout survival and growth. Due to intensive fishing, the lake is stocked on an annual basis.

Black Nugget Pond located near Round Hill, Alberta, was created by the Dodds Coal Mine. The pond was first examined in 1957 for sport fishing potential. The pond has a surface area of approximately 7.3 ha with a maximum water depth of 5.4 m. It was formed by a series of abandoned mine pits, which created irregular lagoons. The 1957 investigation revealed that the concentration of oxygen in surface water was 6.4 mg/l, and the concentration of dissolved solids was greater than 900 mg/l, with unusually high concentrations of sulphates accounting for half of the dissolved solids. Plankton and benthic fauna were scarce. Despite these poor habitat characteristics, the pond was stocked with fish the same year. Since 1970 the pond has been stocked with rainbow trout on an annual basis. Detail chemical analysis of the water conducted in 1985 showed high concentrations of dissolved solids (826 mg/l), sulphate (397 mg/l) and nitrate (1.3 mg/l). The range of the concentrations of cations was also unusual for lakes in Alberta with sodium being the dominant cation (167.4 mg/l) followed by calcium, magnesium and potassium. Other measured parameters were within the natural range occurring in surface waters. Occasionally recorded concentrations of dissolved oxygen varied between 6.2 and 9 mg/l. In the summer of 1986 five rainbow trout measuring approximately 340 to 390 mm in length were submitted for mercury analysis. The results showed that the concentration of mercury in the fish tissue was well below the recommended standard of 0.5 μg/g, and varied between 0.088 and 0.225 μg/g.

Little information is available on the generation of the Pleasure Island Pond near Camrose, Alberta. Alberta Fish and Wildlife recorded that the pond used to support good recreational fisheries on a "put-and-take" basis for rainbow trout. However, stocking of the pond with fish was discontinued in 1979 due a dispute over the pond access between a private land owner and the public.

At Whitewood Mine near Wabamum Lake, Alberta, a lake was created as a replacement for two shallow lakes that had to be drained during the mining operations. Construction for the replacement lake started in 1987, and was scheduled to be completed in 1990. The replacement lake covers 18.5 ha, has a shoreline of 3,500 m and water depth of approximately 7.8 m. The lake site has been designed to have the potential for development as a "put-and-take" fishery, day use recreation and nature viewing. Special features planned to be incorporated in the lake design were a picnic area, campsite, boat launching ramp and beach.

A mine cut near Canmore, Alberta, was successfully developed into a sports fisheries lake. The development of the lake was a part of a joint program conducted by Alberta Environment's Land and Reclamation Project and Alberta Fish and Wildlife Division. Reclamation activities involved re-sloping of steep banks to less than 30° and re-contouring of old spoil pile into the surrounding landscape. A ramp for canoes and rowboats was developed. The lake has a 2 ha surface area, and maximum water depth of approximately 30 m with an average depth of 18 m. The site is extensively used for fishing and picnicking.

The lakes generated in surface coal mining areas in Alberta have unique physico-chemical properties and are located in a mountainous area characterized by restricted climatic and physical conditions. For theses reasons, the Coal Valley Mine and the Luscar Mine, Alberta, conducted a research program to examine and evaluate construction and management guidelines for creating lakes in this environmentally-sensitive region. The program was divided into the following phases:

1. Review of the North American literature pertinent to the development and characterization of lakes and ponds created by surface coal mine operations (survey was conducted in 1991 by Luscar Ltd.).
2. Limnological and fisheries investigation of study lakes (conducted by Luscar Ltd. and Pisces from 1991 to 1993).
3. Fisheries enhancement program implementation and assessment (conducted by Coal Valley Mine from 1992 to1993).
4. Cost-benefit analysis of lake construction and development (conducted by Coal Valley and Luscar mines in 1994).
5. Ecological benefits of lake development (conducted by Bighorn in 1994).

Four lakes were selected for the study: Silkstone Lake, Lovett Lake, Lac Des Roches, and Fairfax Lake. Silkstone and Lovett Lakes are located within the Coal Valley Mine area about 80 km south of Edson, Alberta (Fig. 12). These two lakes, constructed from 1985 to 1986, were the first mine lakes designed specifically for fisheries and recreational uses in the Alberta Foothills. They represent water bodies produced following dragline mining operations. The third lake, Lac Des Roches, is located in the Cardinal River Mine area about 40 km south of Hinton, Alberta. This lake was formed in 1987 in one of the mine's pits and represents water bodies produced by truck and shovel mining. The limnological investigation also included naturally formed Fairfax Lake. This lake, located near the Silkstone and Lovett Lakes, served as a "baseline" and a source of biological material for transplanting and inoculation for these other two lakes.

During the project the following issues were addressed:

1. Characterization of existing habitat conditions in three coal mine lakes created in pits produced by two surface mining methods: dragline, and truck and shovel, to establish specific considerations for the region.
2. Evaluation of various habitat enhancement alternatives to maintain or improve the characteristics of surface mine lakes for fisheries habitat.
3. Examination and establishment of pit lake development and management guidelines for optimum management practices.
4. Documentation of the benefits of lake development to wildlife and the areas ecosystem.
5. Analysis of economic benefits of lake development.

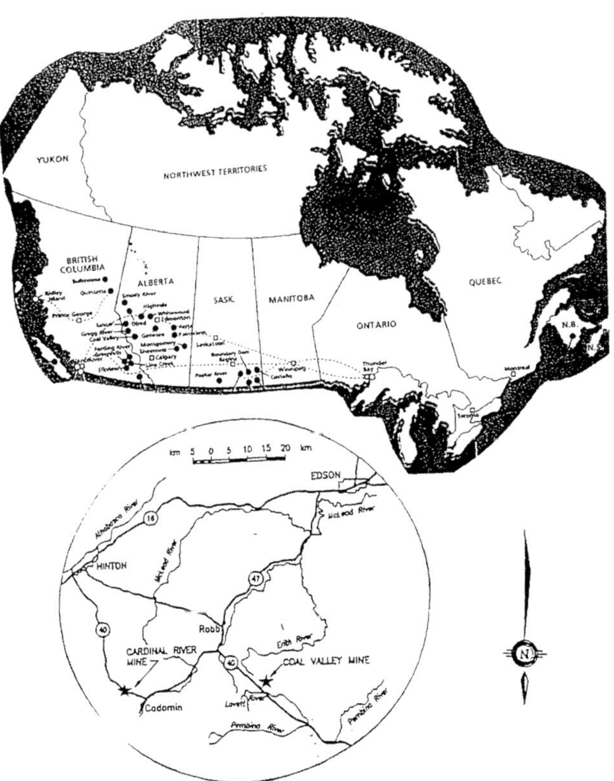

Fig. 12. Location of the Coal Valley, Alberta, Canada

5.1 Construction of the Lakes

The initial development of Silkstone and Lovett Lakes involved re-sloping and leveling operations on the shore line and bottom configuration, top soil replacement and seeding of the surrounding area. The construction of the lakes was completed and the final, stable water levels were achieved in 1985 and 1986 for Lovett Lake and Silkstone Lake, respectively. Both lakes are dependent on inflow from groundwater sources and runoff from their surrounding drainage basins to maintain lake levels. Silkstone Lake has an outflow to the nearby Lovett River, with exit flows dependent upon lakes levels. In the case of Lovett Lake, underground seepage out of the lake maintains the shore line at a constant level. Littoral zones (i.e., area less than 3 m deep) were constructed to account for over 30 % of the lake areas. The shoreline and bottom configuration were left irregular to increase habitat diversity and minimize wave action. Additional development activities involved introduction of macrophytes, transplanted from Fairfax Lake and an isolated ox-bow of the Lovett River located in close proximity to the mine site. Physical and hydrological characteristics of the lakes are summarized in Table 1. The morphometry of Silkstone and Lovett Lakes is shown in Fig. 13, and Fig. 14 is an areal view of both lakes.

Table 1. Physical and Hydrological Characteristics of Lac Des Roches and Lovett, Silkstone and Fairfax Lakes

Parameter	Lovett Lake	Silkstone Lake	Fairfax Lake	Lac Des Roches
Surface Area [ha]	6.0	6.4	28.4	16.2
Maximum Depth [m]	18.0	14.8	7.6	70
Mean Depth [m]	5.5	4.7	3.2	37
Volume [m^3]	330,000	300,800	909	6,000,000
Watershed Area [ha]	161.6	146.2	153.7	No information
Outflow Channel	No	Yes	Yes	Yes
Shoreline Length [km]	1.34	1.35	No information	1.98
Littoral Zone [%]	32	37	63	5
Elevation [m]	1,360	1,360	1,360	1,593

The third study lake, Lac Des Roches, was created by Cardinal River Coals, Ltd. in an open pit formed by truck and shovel operations. The decision to create the lake originated from the prohibitive reclamation costs associated with re-handling sufficient overburden for backfilling of the pit, according to the governmental specifications. As the pit was located in a natural creek valley and water supply could be guaranteed, the approval from the government agency was granted (Luscar Ltd. et al. 1994). The development of the lake was initiated in 1985. It involved construction of 0.7 ha of littoral zone, re-sloping of overburden, placement of top soil and re-vegetation of the surrounding area. The majority of the littoral area of the lake was constructed at the east end, where the lake water exits to the natural channel of Jarvis Creek West. A berm was built at the final water elevation, to separate the warmer, more productive shallow area from the main body of the lake. The berm also serves to intercept heavy wave action during windy periods. The littoral zone area represents only approximately 5 % of the total lake surface area. Any further development of the littoral zone would have involved placement of backfill at a prohibitive cost. The remedial work focused, therefore, on providing accessible shoreline and habitat enhancement. Physical and hydrological characteristics of the lake are summarized in Table 1. The morphometry of the lake is shown in Fig. 15, and Fig. 16 is an areal view of Lac Des Roches.

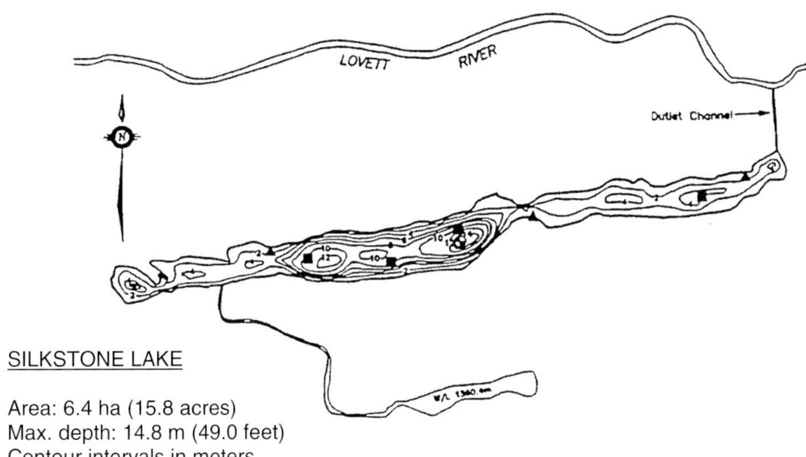

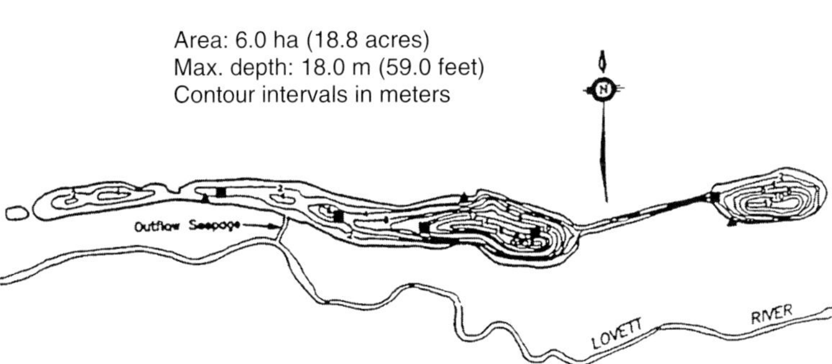

Fig. 13. Morphometry of the Silkstone and Lovett Lakes

A number of measures were taken to promote wildlife at Lac Des Roches. A large number of truck tires were placed in the shallow safety bench to offer an area for periphyton attachment and fish cover. Organic soil and partially decomposed hay were placed in the littoral zone to provide suitable habitat for benthic invertebrate and macrophyte communities. Macrophytes from nearby Mary Gregg Lake were transplanted. Reclamation activities also included seeding of high walls and the construction of benches for bighorn sheep and mule deer which regularly utilize the lake area (Luscar Ltd. et al. 1994).

The lake development program was finalized in 1987 when a full water supply level was reached. The source of water included ground water, the diverted flow from upstream Luscar Lake and surface runoff.

Fig. 14. Aerial view of Silkstone and Lovett Lakes

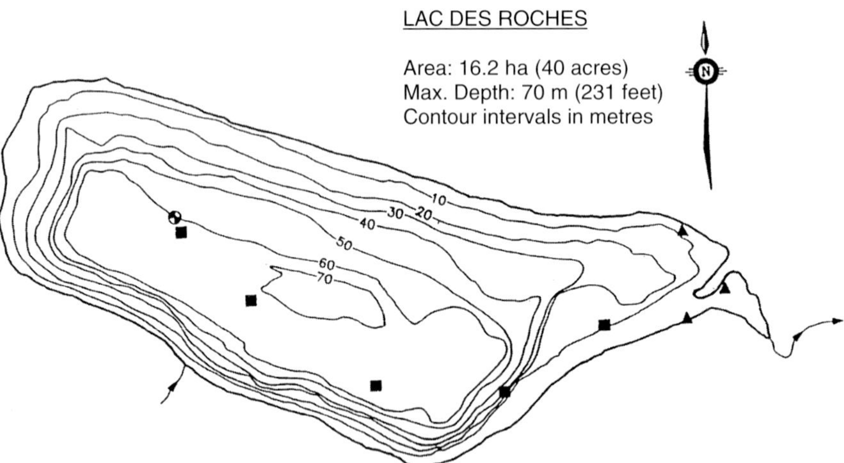

Fig. 15. Morphometry of Lac des Roches

Fig. 16. Aerial view of Lac des Roches

An extensive habitat improvement program of the lake outflow channel was undertaken in 1987 to improve the spawning potential immediately downstream of the lake. The program included the construction of a gabion mat and a 4-m deep plunge pool to ensure a constant water level. Small step dams placed at 8-m intervals over a 75-m distance created numerous small pools. Fine washed gravel was placed behind each dam to provide spawning areas. Rainbow and brook trout were observed in these areas since 1988 (Luscar Ltd. et al. 1994).

5.2 Results of Lake Studies

Results of the limnological investigation indicated that the lakes are capable of supporting fish populations suitable for sport fishing. Initially, the water in all three generated lakes was turbid. However, it reached an excellent clarity within three years, and it continues to exhibit very low turbidity. The concentrations of compounds potentially toxic to aquatic life, such as metals, phenols, nitrites and cyanides, are very low and remained below the federal and provincial freshwater quality guidelines and objectives.

Several water quality indicators, such as sodium, sulphate, alkalinity, bicarbonate, and dissolved solids, are present at greater concentrations in the pit lakes than in the naturally formed lakes in the area (i.e., Fairfax Lake). However, the levels of these parameters are within the natural variation for surface waters and are not high enough to adversely affect aquatic life. Elevated levels of these parameters are expected in areas filled with groundwater. Nitrites, typically found at high concentrations in mine pit lakes, have no apparent effects on the primary productivity of the lakes.

Based on their chemical and biological characteristics, Lovett and Silkstone Lakes are oligo-mesotrophic. Both lakes have a high density and diversity of benthic invertebrates, comparable to those observed in Fairfax Lake. These characteristics remained unchanged after the introduction of fish to the lakes in 1991 and 1992. The main impact of fish predation appears to be a decrease in the number of zooplankton larger than 1 mm. This is common in lakes following the introduction of fish (Luscar, Ltd. et al. 1994).

Lac Des Roches is oligotrophic. The low productivity of the lake is regulated to a large extent by its morphometric characteristics, i.e., the relatively large size and water depth, with limited mixing of the deeper part of the lake, prevents recycling of nutrients. Another factor limiting the lake productivity appears to be the relatively small littoral zone. There was no substantial change in chemical and biological character of the lake due to fish predation over the 1991 to 1993 period of investigation.

The studies showed that the three mine lakes, Lovelett and Silkstone Lakes and Lac Des Roches, can support sport fish population and that the growth rates of the fish are superior to those in adjacent natural water bodies. Lovelett and Silkstone Lakes have been approaching maturity and stability with aquatic communities that are similar to those in the neighbouring natural lakes. Both lakes are productive, with abundant and diverse invertebrate populations that are able to sustain high fish growth rates. Both lakes were stocked with hatchery rainbow trout in the spring of 1991 and 1992. The follow-up fisheries investigation revealed that the fish growth in the lakes exceeded that of the natural lake in the immediate vicinity. This was due to the abundance in food organisms that had accumulated in the 4- to 6-year period prior to fish introduction. However, both lakes lacked escape cover for fish at the time of stocking and fish were susceptible to predation, which considerably reduced survival rates. The low survival rate of fish in Silkstone Lake was additionally caused by fish escaping through the outflow channel. These problems were mitigated in 1993 by introducing additional escape cover (i.e., rock and brush piles) and modifying the outflow channel. Maintenance stocking of rainbow trout will be required in both Silkstone and Lovelett Lakes.

Lac Des Roches is unique to Alberta because it may be the first created mountain lake that has become occupied by native fish from downstream. The lake is relatively cold and deep with a small littoral zone relative to its area. These features make the lake an oligotrophic water body with limited fish production capabilities. However, the lake approaches the ideal objective - it has become occupied by native fish and is a self-sustaining recreational sport fishery lake. Two native fish species occupy the lake: rainbow trout and bull trout. Both species are of particular interest. Bull trout are a species of concern in North America and the Athabasca strain of rainbow trout has recently been noted as a unique population with possible future subspecific status. Enhancement of spawning habitat in the outlet stream from the lake has been effective and it is intensively used by rainbow trout.

All three lakes are a valuable addition to the lake ecosystem and to the diversity of fish habitats in the eastern Slopes of Alberta, particularly for their ability to support self reproducing populations of native species. These man-made features add new waterbodies to a region that is not typically characterized by standing water. The addition of new waterbodies to the region has augmented or provided habitats for many species. Generally, the lakes provide staging areas for migrant shore birds and waterfowl, with feeding on invertebrates on the route to the Arctic. Further, the lakes provide breeding habitats for local amphibians, which establish another link in the food chain. The lakes are a source of water for many animals living in the area, such as elk, mule deer, white-tailed deer and moose, and are habitat for aquatic fur-bearers, such as muskrats and mink, as well as for marsh birds.

Eighteen new species of water-oriented birds, which had not been observed prior to the development of the lakes, were noted during the period of the study of the lakes. Occurrence of many of the birds was attributed largely to the presence of the man-made lakes.

5.3 Costs of Lake Developments

The reclamation of coal mine dragline pits and truck/shovel pits to lake habitat has proven to be a desirable economic alternative to upland forest reclamation. The cost savings are associated with less backfilling and earthmoving costs than would be required to reclaim the mining area to accepted standards of upland forest. The cost of a lake development can be relatively consistent from one area to the other. However, due to the amount of backfilling and sloping/leveling required, the cost of generating dry land is quite variable. Mining conditions that affect the relative reclamation cost, i.e., dry land vs. lake, are: thickness of coal seam, down-dip angle of coal, depth of overburden, and width of final cut.

The cost analysis conducted by the Coal Valley Mine in Alberta, which compared the lake development in dragline operations to upland forest reclamation, showed that the savings ranged between Can \$50,000/ha and Can \$135,000/ha or approximately Can \$0.85 per clean metric tone (CMT). The cost analysis conducted by the Luscar Mine, Alberta, showed that the lake development in truck/shovel operations in mountainous settings may result in savings ranging between Can \$151,000/ha and Can \$416,000/ha or Can \$0.80 to Can \$3.00 per CMT, depending on lake configuration and coal recovery.

5.4 Management of the Lakes

The study of the three mine lakes contributed significantly to the development of a working model for post-mining land use that will have applications to other mining operations in Alberta's eastern slope region. The results of the study will be used to improve the recreational potential of the area. From the fishery management perspective, the following components need to be evaluated and implemented in the development of further lakes: lake water supply (hydrology), which must be considered in the pre-construction planning stage; watershed reclamation, erosion control and basin stabilization activities; lake morphometry and habitat enhancement; post construction lake monitoring and management.

The results of the habitat investigation and literature review suggested that there are many critical habitat components that must be considered when establishing viable sport fisheries in mine pit lakes. Some of these components, such as size and depth of the lake and orientation of the lake, will be predetermined by excavation practices and other mining operations. However, many important factors in lake habitat design, such as shoreline contour and slope, litoral zone, substrate type and composition, can be manipulated. The most important factor appears to be the establishment of adequate littoral zone, provision of escape cover for fish, introduction of invertebrates and macrophytes from local sources, and the delayed introduction of fish to allow biological communities in the lake to approach stability. It takes approximately 2 to 5 years to establish aquatic vegetation after the lake is filled with water. The establishment of aquatic plants in lakes is ecologically important. The plants form the base of a complex food web and have a major impact on wildlife by providing food and cover for wetland mammals, such as muskrats, nesting structures for marsh birds, food and cover for aquatic invertebrates, amphibians and fish, cover for nesting waterfowl and food for migrating waterfowl. It is desirable to promote self reproducing populations and avoid the potentially expensive long term commitment associated with fish stocking program. However, this may only be possible in lakes with features similar to those of Lac Des Roches.

6 Overview of Opencast Coal Mining and Reclamation in the UK

The UK has a long tradition of coal mining and although overall production has decreased in the past half-century, there has been an increase in opencast mining since 1970. The main regions of shallow coal are the Scottish lowlands, the Northeast of England, the central East (including Yorkshire and Nottinghamshire), the Northwest (Cumbria, Lancashire and North Wales), the Midlands and South Wales (Fig. 17). Opencast mining in these areas is of particular environmental concern due to the high population densities. Furthermore, since UK surface mines are generally small by world standards, averaging only 200 hectares in surface area, site lives average 5 to 6 years only and consequently 10 to 12 new pits must be opened each year to maintain current capacity.

Opencast mining began in 1942 to meet the War demand and reached a local production peak of 15 million tons per year prior to the 1958 Opencast Coal Act. With the shift in responsibility to local planning authorities, production dropped, until in 1974 the Plan for Coal, sparked by the Oil Crisis, set a target of 15 million tons again. This target was reached by 1982. Despite a recommendation by the Flowers commission in 1981 for a reduction in opencast mining, by 1992 the production reached 19 million tons.

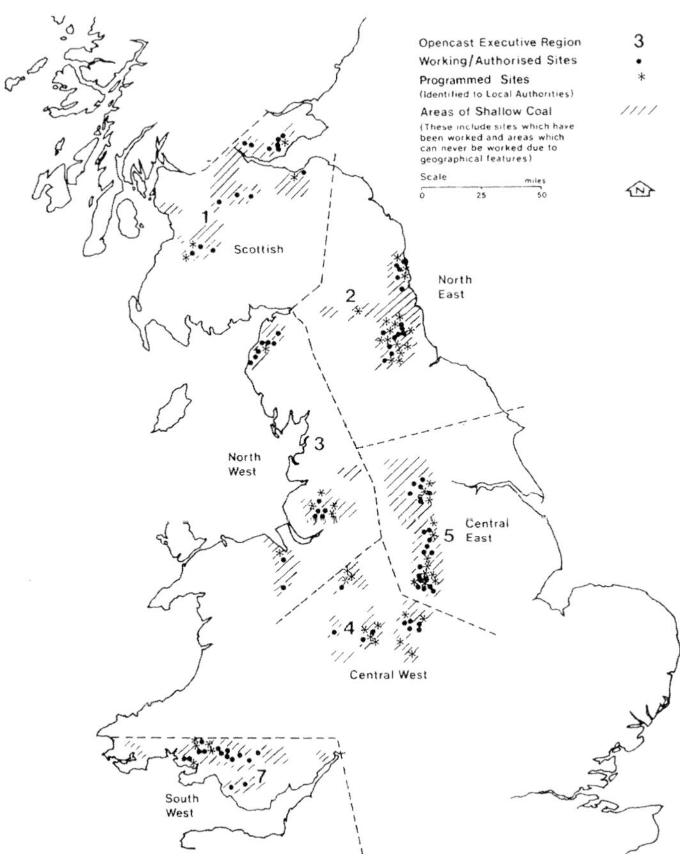

Fig. 17. Areas of present and potential opencast working in the UK (HMSO 1981)

Relative to total coal output, the increase in opencast activity is quite dramatic. From 1965 to 1985 the percentage of opencast coal increased from 5 % to 30 %. This increase can be attributed to the profitability of opencast coal during a period that saw coal subsidies reduced and market forces encouraged. The miners strike of the early 1980s may also have been important, since it only involved workers from underground mines.

Opencast lands have been predominantly restored for agricultural use in the UK, although there has been a shift away from this policy in recent years (Proctor 1990). As UK surface mines have a high ratio of overburden to coal (on average 19 to 1) it is possible to restore most areas close to their original contours. From 1942 to 1990 of the 60,000 acres of worked land, 90 % was restored for agriculture. However, since the late 1980s there has been a change in practice away from agricultural restoration due to massive European food surpluses and perhaps also criticism of environmental impacts of capital intensive farming techniques in the UK. In many cases, restoration of mining areas to the natural environment is now considered. In the early 90s hedgerows were being planted at the rate of 40 km/year on former mining lands. Habitats are now being designed with woodlands, ponds, ditches, wetlands and sanctuaries for birds and wildlife. Proctor (1990) reports on the return of a river to its natural sinuous path and the creation of heather moorland in a former upland mining area.

Alternative reclamation strategies now include the formation of leisure areas, country parks, golf courses and ski slopes, as well as refuse disposal sites. At one site near Shipley, an area of former industrial dereliction has been transformed, post surface mining, to a commercial theme park and country park. The country park includes a series of lakes for sailing, windsurfing, water-skiing, fishing and wildlife (Brook 1989).

A proposed development scheme for the Windsor opencast mine in West Yorkshire highlights the attention to nature conservation that is considered in UK mine reclamation (R.J.B. Mining U.K. Ltd. 1997). While the plan is to recreate a sustainable agricultural landscape, which contributes to the rural economy, the development includes:

- an increase in species and habitat diversity,
- the accommodation of locally occurring protected species,
- an increase in open and running water habitat,
- a doubling of woodland cover in the district,
- the protection and enhancement of green corridors,
- an increase in public rights-of-way to encourage countryside recreation.

The Windsor restoration scheme includes the planting of at least six categories of woodland, four categories of grasses, four categories of herbs and further categories of hedgerow plants and reeds, with five to ten different species per category.

References

Baker RA, Wildshire AG (1970) Microbial factor in acid mine drainage formation: a pilot plant study. Env Science and Technol 4/5:401–407

Brook C (1989) Environmental aspects of opencast mining with reference to the Rother Valley Country Park. Mining Engineer:191–196

Grecula P. et al. (1997) Mineral resources of Slovakia, Min. of the Environ. of Slovak Republic, Geolog. Survey of Slovak Republic

Guziel A (1988) Ochrona i kształtowanie środowiska w rozwoju górnictwa w Polsce (Protection and environmental formation in development of mining in Poland) SGGW-AR Warszawa Część I (Part I) (in Polish) pp 50–66

HMSO (1981) Coal and the Environment, Commission on Energy and the Environment, HM Stationary Office, London

Jędrczak A (1992) Skład chemiczny wód pojezierza antrpogenicznego w Łuku Mużakowskim (Chemical composition of water of anthropogenic lake district in the "Łuk Mużakowski). Zielona Góra. WSI (in Polish with English sum.) pp 54, 140

Koprowska L (1995) Sukcesja fauny dennej w zbiornikach powstałych po wydobyciu węgla brunatnego (Successions of bottom fauna in reservoirs arisen after postexploitation of lignite). AR-T Olsztyn. doctor thesis (in Polish) 42 p

Kozacki L (1976) Jeziora antropogeniczne, ich znaczenie i możliwości zagospodarowania (Anthropogenic lakes, their significance and possibilities of management). Jeziora Ziemi Lubuskiej ich wykorzystanie i ochrona przed zanieczyszczeniami (Lakes in the Ziemia Lubuska their use and protection against pollution). Proceedings Symposium Naukowe, Łagów 18-19.05. 1976. Zielona Góra (in Polish) pp 141–150

Luscar Ltd., Luscar-Sterco Ltd., Cardinal River Coals Ltd., Pisces Environmental Consulting Services Ltd., Bighorn Environmental Design Ltd. (1994) Development of sport fisheries in lakes created by coal mining operations in the Eastern slopes, Luscar Ltd., Cardinal River Coals, Ltd., Pisces Environmental Consulting Services Ltd., and Bighorn Environmental Design, Ltd., Edmonton, Alberta, Canada, 152 p

Matejczuk W (1986) Charakterystyka zbiorników wodnych w wyrobiskach poeksploatacyjnych węgla brunatnego. (Characteristics of water reservoirs in postexploitation workings of lignite). Politechnika Wrocławska, Instytut Inżynierii Ochrony Środowiska. Wrocław (in Polish)

Matejczuk W (1989) Plankton poeksploatacyjnych zbiorników wodnych z rejonu Trzebiela (Plankton of postexploitation water reservoirs in the Trzebiel region) Przyroda Środkowego Nadodrza (Nature of Odra middle course). Zielona Góra (in Polish) pp 92–118

Pietsch W (1970) Ökophysiologische Untersuchungen an Tagebaugewässern der Lausitz. Habillitationschrift, Dresden

Proctor R (1990) Opencast Restoration in the UK, article from 'Minetech 90' reprinted in Quarry Management

R.J.B. Mining U.K. Ltd. (1997) Windsor opencast coal site: Proposed development

Ryhänen R (1961) Über die Einwirkung von Grubenabfällen auf einen dystrophen See. Ann Zool Soc "Vrząd" 22/8:1–70

Statistical yearbook (Rocznik Statystyczny) (1997) Główny Urząd Statystyczny, Warszawa (in Polish with English and Russian List of Tables) 715 pp

Stottmeister U, Gläßer W, Klapper H, Weißbrodt E, Eccarius B, Kennedy C, Schultze M, Wendt-Potthoff K, Frömmichen R, Schreck P, Strauch G (1999) Strategies for Remediation of Former Opencast Mining Areas in Eastern Germany, Chapt. 16 In: Azcue JM (ed) Environmental Impacts of Mining Activities. Springer Berlin Heidelberg New York

Šucha V et al. (1995) Determination of the degree and extent of the countryside destruction as a result of ecological breakdown at the locality Banská Štiavnica – Šobov and estimation of the possibilities of restoration

Šucha V et al. (1996) Complex model of environmental effects as a result of raw material mining in typical areas in Slovakia. Particular final report. Manuscript. Geofond Bratislava. 118 p

Svoboda I (1993) Land Restoration in the North Czech Brown Coal Basin: the new challenges and opportunities. Int J Environ: Issues in Minerals and Energy Industry:151–154

Alternatives for the Reclamation of Surface Mined Lands

Christopher Kennedy

University of Toronto, Department of Civil Engineering, 35 St. George Street, Toronto, Ontario, Canada M5S 1A4

While the main focus of this report is the formation of lakes from former surface mines, this chapter briefly considers other alternatives for surface mine reclamation. The reasons for discussing other after uses, such as forestry, agriculture, natural wildlife and recreation, are twofold. Firstly, formation of lakes from former mine excavations will not always be the best strategy, and may even be difficult under some geological conditions. Secondly, in cases where the generation of lakes is the chosen approach, consideration will often still be required for impacted lands immediately surrounding the water body. In many existing reclamation schemes a mix of alternatives is chosen.

Reclamation of surface mined lands has been undertaken for over a century or more; however, it would seem that over the last 20 or so years there has been increased interest in reclamation strategies. Richardson (1977) notes that tree plantations were used to restore acid shale tips in Cumbria, England in 1898. Forestation of mine spoils in North America started in Indiana in 1928 (Medvick 1980). Increasing concern over restoration practices in recent decades has resulted in tougher regulations, such as the U.S. Surface Mining and Reclamation Act of 1977. Consequently, methods of surface mine reclamation continue to be studied and developed.

Given the large amount of previous work on the subject of reclamation, much of which is undertaken by the mining industry itself, no attempt will be made here to cover the topic in its entirety. Rather, an overview of surface reclamation activity will be given, with slightly more emphasis on practical methodology, as opposed to scientific principles. This chapter draws heavily upon a two-volume text on the subject edited by Hossner (1988), which is a good general source for further information.

1 Climate

The first consideration in any reclamation plan is the climate of the impacted region (U.S. NRC 1981). Precipitation, evapotranspiration, air temperatures and the length of the growing season are particularly important for the establishment of a new environment. Sufficient moisture is required for the establishment and maintenance of vegetation. The timing of rainfall should also be a consideration; it is clearly beneficial that the majority of annual precipitation occurs during the growing season. The frequency and intensity of storms also affect surface run-off and consequently the potential for soil erosion. Air temperatures, and in particular annual extreme values, influence the rate of plant growth, species formation and the rate of soil formation. Plant species diversity and the rate of soil formation will generally be greater at sites with long growing seasons.

While a choice of vegetation native to the region of the reclaimed lands should clearly alleviate concerns over suitability for the climate, consideration should also be given to microclimate effects. Air temperatures close to the ground surface are influenced by soil colour, surface aspect relative to the sun, soil moisture content and shading by existing plant communities (Hadley 1961).

2 Soil Reconstruction

Soil properties are of importance to possibly all reclamation alternatives, so some general discussion of soil reconstruction may be given here, before specifically focussing on forestry, agriculture, and wildlife environments. Post-mining establishment and growth of plants is often more limited by physical aspects of soil, rather than chemical problems (Dollhopf and Postle 1988); thus, the discussion here will give more emphasis to the former. Common physical problems include erosion, crusting, infiltration, compaction, excessive swelling and structural degradation. Generally speaking it is easier to correct for chemical defects, e.g. by the addition of fertilizer, than overcome physical soil problems. However, this generality should be clarified. The objective of reclamation is to create an environment that in the long run does not depend on intensive management in order to maintain a stable landscape (U.S. NRC 1981). A reclamation scheme that requires long-term chemical applications is therefore inadequate.

Arguably the primary physical problem in soil reconstruction is overcoming the effects of compaction. Soil compaction impedes plant growth by restricting water movement, nutrient intake and oxygen; accumulating carbon dioxide; and hindering root development. Further discussion of these problems by Kendle and Schofield (1992) is summarized in Table 1. A particularly crucial consequence of soil compaction is that it results in reduced infiltration capacity, leading to lower water availability, and increased run-off with associated problems of soil erosion. One of the greatest challenges of reclamation is to reestablish a landscape with a sustainable infiltration runoff relationship (U.S. NRC 1981).

One measure of the effect of soil compaction on root growth is the critical bulk density of soil at which root penetration is prevented or severely restricted. Table 2 provides a summary of critical bulk densities for non-mined soils reported by Barnhisel (1988); the data comes from several sources with differing experimental conditions.

Table 1. Effects of soil compaction (based on Kendle and Schofield 1992)

Increased mechanical impedance	Inhibits the ability of roots to push through the soil
Reduced aeration	Larger soil pores are compressed, restricting the movement of water. Oxygen takes longer to enter the soil and carbon dioxide escapes more slowly
Changes in moisture availability	Water is gradually forced from between soil particles. Remaining water is in smaller pores where it is bound tightly to soil particles making it unavailable to plants. Slower flow paths through the soil causes reduced infiltration. Run-off is greater over compacted soil potentially causing erosion of top-soil.
Changes in thermal conductivity	Closer packing of soil particles may give rise to increased thermal conductivity. This has both beneficial and detrimental potential implications
Changes in nutrient availability	Changes in moisture level and oxygen content may effect nutrient availability in several ways: • reduced absorption of nutrients by roots • less nitrogen mineralized from organic matter • increase the exchange of nutrient ions with solid particles • change the valence state of iron or manganese potentially leading to toxic levels of these metals • reduce sulphur to phytotoxic hydrogen sulphide • increase the solubilisation of soil phosphate • cause the anaerobic decomposition of soil nitrate leading to loss of nitrogen
Microbial population changes	Compaction can greatly reduce microbial activity, e.g. inhibiting the mineralisation of soil nitrogen. Some pathogenic diseases are more aggressive on wet, poorly drained soils.

Table 2. Effect of bulk density on root penetration for various plant species (Barnhisel 1988)

Plant species	Critical bulk density	Soil texture
Alfalfa	1.75	Sandy loam
Corn	1.69-1.80	Silty clay
Corn	1.80	Silty clay loam
Corn	1.67	Sandy clay loam
Corn		1.901 Sandy clay loam
Corn	1.80	Silty clay
Cotton	1.70	Sandy clay loam
Cotton	1.87	Sandy loam
Cotton	1.78	Sandy loam
Cotton	1.88	Sandy loam
Grain sorghum	1.60	Silt loam
Soybeans	1.60	Silty clay loam
Sunflower	1.75	Sands
Sunflower	1.46-1.63	Clays
Wheat	1.60	Silt loam

In addition to bulk density, several other soil properties may be measured to determine the suitability of overburden material for use as topsoil or subsoil: particle size distribution, rock fragments, clay mineralogy, hydraulic conductivity or infiltration rate, plant available water holding capacity and erodibility potential (Dollhopf and Postle 1988).

Suitable values for infiltration rates and plant available water are given in Tables 3 and 4 Two common methods of measuring infiltration rates are the use of ring infiltrometers and rainfall simulators (Bertrand 1965; Dingman 1994). Plant available water holding capacity is determined by desorbing the soil in a pressure plate apparatus (Richards 1965).

Table 3. Classification of infiltration rates (U.S. Soil Conservation Service 1951)

Descriptive term	Infiltration rate [mm/hr]
Very rapid	> 254
Rapid	127 – 254
Moderately rapid	63 – 127
Moderate	20 – 63
Moderately slow	5 – 20
Slow	1 – 5
Very slow	< 1

Table 4. Rating of plant available water-holding capacities for various moisture regimes potential (Dollhopf and Postle 1988)

Rating	Aquic and perudic [cm/100 cm]	Udic and ustic [cm/150 cm]	Aridic and xeric [cm/150 cm]
Very low	< 5	< 7.6	< 6.4
Low	5.1 – 7.6	7.6 – 15.2	6.4 – 12.7
Moderate	7.6 – 10.2	15.2 – 22.9	12.7 – 19.0
High	> 10.2	22.9 – 30.5	19.0 – 25.4
Very high		> 30.5	> 25.4

Topsoil for crop use should ideally consist of medium-grained materials such as silt loams, loams or silty clay loams (Jansen and Melsted 1988). Clays and organic matter provide nutrients and stabilize soil structure, but too much of these materials will give poor tilth, low conductivity and poor aeration. In particular, soils with a high content of smectite clay may be prone to shrink-swell, cracking, crusting, high saturation percentages and reduced infiltration; these effects are further increased by the presence of sodium salts (Dollhopf and Postle 1988). Jansen and Melsted (1988) suggest that 20 to 30 % clay or organic content is suitable for the surface and perhaps 20 to 35 % for the subsurface. Soils that are low in organic matter may seal over during heavy precipitation, resulting in low infiltration and crusting upon drying.

Rock fragments are not necessarily undesirable in all reclamation schemes. Many types of rock may hold water at low tension, thus making it more easily available for plant use (Ashby et al. 1984). Rock fragments may additionally reduce evaporation, alter soil temperatures and store additional nutrients (Lutz 1952). Nevertheless, coarse materials do cause difficulties if the end-use of the land is agricultural and requires tillage or mowing.

In well-planned mining operations it is usual to save and store separately the A-horizon of the topsoil when it is rich in organic matter and microorganisms. Placement of A-soil over less nutritious mine spoils yields obvious benefits. Studies in North Dakota showed that vegetation yields increased in proportion to the depth of the topsoil up to a depth of 75 cm (Power 1978). Mixing of A-horizon with sub-soil overburden dilutes beneficial effects.

There may be benefits to segregating and storing other original soil layers. For instance when the A-horizon is too thin, or contains toxic material, it may be worth stockpiling a fertile B-horizon. In areas of the Indiana coal region, deeper Lacustrine sediments were found with organic matter comparable to or surpassing that in the A-horizon (Byrnes et al 1980).

Following the identification of suitable soils, the recontouring of the landscape is a key component of the soil reconstruction process. In addition to improving the aesthetic appearance of land, the primary objective of recontouring is to establish a landscape with a stable infiltration runoff relationship, thereby reducing the potential for erosion. This involves the control of ponding and encouragement of natural drainage. Further objectives of recontouring may be to eliminate landslide potential, control water pollution and eliminate hazards such as high cliffs and deep pits (U.S. EPA 1973).

The recontouring process can in cases improve the utility of the landscape beyond its pre-mining condition. For example grading can remove obstacles and create more favorable slopes for agricultural machinery. Nevertheless, in some countries legislation dictates that landscapes should be returned as near as possible to original contours.

A number of recontouring strategies can be used to combat increased runoff rates, which result from compaction and breakdown of soil structure. One approach is to reduce slopes below their pre-mining gradients to make water flow more slowly over the surface. This strategy was used in the Rhine region of Germany, where slopes were reduced to a maximum gradient of 1.5 % (Heide 1973). Experiments on surface configurations of mined lands by Dollhopf and Goering (1982) found that large surface depressions created by dozers could reduce runoff by 72 % and slope erosion by 92 % in comparison with more conventional chiseling treatment. Furthermore, microtopographic preparations such as chiseling and gouging tend to only be effective at runoff and erosion control for short periods of a year or two before filling with sediment. Dozer basins may be effective from 10 to 50 years. Other runoff control strategies include contoured terraces, furrows and trenches (U.S. NRC 1981; Verma and Thames 1978). Further discussion of the design of post-mining landscapes is given by Schaefer et al. (1980) and Ventura and Dougherty (1980). Computer software can aid in the recontouring design, to reduce costs and assess the aesthetics of a new landscape prior to regrading (Nicholson 1995; Russell 1996).

The type of machinery used in reclamation contouring can have considerable impact upon the extent of soil compaction. Rubber tired earth moving scrapers are poor in this respect. Rear loading dump trucks cause less compaction due to their lower dead weight. Dump trucks do require a separate loading system, but can be more efficient on long hauls. Despite their low ground pressure, dozers will cause compaction in any slightly moist soils. This is because of their large dead weight, compounded by vibration from the dozer tracks and the compressive force applied by the blade when pushing soil. Further discussion of spoil handling techniques is given by Harwood and Thames (1988).

3 Forestry

There are several reasons why forestry might be chosen in a surface mine reclamation scheme: slopes may be too steep for agricultural use; the production of timber gives an economic benefit; or trees might be planted simply for aesthetic reasons.

Once an area has been designated for forestry, two important considerations are whether or not grading of the surface and decompaction are required. As discussed in the previous section, excessive grading of mined lands causes increased compaction and reduced infiltration, leading to poor productivity in vegetation. If a site is generally not toxic to trees and appropriate amounts of nutrients (nitrogen, phosphorus, potassium) are present or can be applied, then survival rates and growth rates for trees may be greater if they are planted on non-graded spoil banks (Powell 1988). Slopes of the soil banks may well be steep, but intermittent depressions will prevent transport of eroded material. Steep banks and depressions also provide shelter to protect vulnerable species from sun and wind in harsh climates (Plass and Powell 1988). If grading can be avoided then a considerable cost saving is also achieved.

Where soils are too compacted for successful forestation, then deep ripping or other forms of decompaction are required. Under such circumstances re-grading may be required in order to create access for decompaction machinery. Powell (1988) suggests that it may be better to perform riping after the placement of top-soil, since the process of placing top-soil may itself cause yet further compaction. Furthermore, any rocks brought to the surface by ripping do not generally impact upon tree growth. However, in the proposed reclamation scheme for a site in Wales the problem of recompaction is avoided by well-organized placement of topsoil (Celtic Energy Ltd. 1997). Following ripping of the overburden, topsoil is to be placed in strips with delivery from the back of a truck. Two excavators, one mounted with a power fork, are used to spread the material and remove large stones. Neither trucks nor excavators pass over the finished surface.

A further technique for decompaction is the use of small explosive charges to create cylindrical holes of up to 50 cm diameter and 75 cm depth, suitable for the placement of a single tree (Richardson 1977).

The choice of tree or shrub species depends upon a number of factors, such as geographic and topographic location, climate, soil characteristics, land management objective, species availability and legal restrictions (Plass and Powell 1988). Considerable knowledge exists with respect to the geographic and climatic zones of species adaptability. Microclimate is also important with greater species selection possible on cooler slopes facing away from the sun. The land management objective may be to create a pleasant woodland with "attractive growth forms, unusual foliage, showy flowers and interesting fruiting bodies" (Davidson 1977); or it may be to create a commercial plantation. Characteristics of several tree and shrub species that are useful to wildlife are given in Table 8. Legal restrictions in some areas may prohibit the choice of anything but native species.

Soil chemistry, and in particular pH, is important in species selection. Tables 5a and b list trees and shrubs that are known to tolerate acid and alkaline spoil material respectively. Powell (1988) notes that the yellow poplar, northern red oak, sycamore, river birch, Norway maple, red maple, sugar maple, cottonwood, bigtooth aspen and quaking aspen will only grow on acid material when no moisture deficiencies exist during the growing season. Several tree species are able to tolerate both acid and alkaline conditions. The European black alder (*Alnus glutinosa*) is particularly hardy and is ideal as a nurse species; it grows rapidly, produces abundant leaf litter and it tolerates a wide range of pH (Richardson 1977).

To establish an attractive mixed forest suitable for wildlife, the species selection should contain a high proportion of hardy, soil–building nurse trees, such as black locust, alder, birch, shrub lespedezas and autumn olive (Powell 1988). Once these trees are established other natural forest vegetation will invade from nearby non-disturbed woodlands. It may take several decades for a diverse well-balanced "natural" forest to form, but in the interim the site will remain stable. If the area is very large and other seed sources are remote, then placement of a few tree spades in strategic locations is desirable. Tree spades should contain a few hardwoods such as oaks, maples and cherry, which are close to seed-bearing age. Soil moisture conditions should also be considered in species selection, with, for example, willows and poplars chosen for wetter areas.

For commercial tree plantations, choice of a single pine or conifer species makes harvesting easier and satisfies market requirements (Plass and Powell 1988). However, to reduce the potential impact of disease or insect damage it may be less risky to establish three or four species with similar growth rates (Powell 1988).

A variety of techniques may be used to plant seedlings or seeds (Plass and Powell 1988). If the terrain is accessible and rock-free, then a mechanical tractor drawn machine can be used to plant seedlings; otherwise hand planting is normal. The seedlings may be planted in growth containers to supply nutrients and minimize damage to roots. In areas with predictable rainfall in the growing season, poplars and willows have been successfully grown from cuttings. An efficient alternative to planting seedlings is to use direct seeding. Seeds may be planted by hand, from a tractor or even by helicopter or fixed-wing aircraft for large areas. Care should be taken when handling seedlings prior to planting, as plants are vulnerable to drying out and other damage while out of the soil (Richardson 1977). In areas of low annual precipitation, such as parts of western USA, new seedlings are sometimes irrigated for the first year using sprinkler systems.

Table 5a. Trees and shrubs known to tolerate acid spoil material (Richardson 1977)

Trees	
Acacia baileyana	acacia
A. melanoxylon	blackwood acacia
Acer negundo	box elder
A. platanoides	Norway maple
A. platanus	American sycamore
A. rubrum	red maple

Table 5.a (continued)

A. saccharum	sugar maple
Alnus glutionosa	European black alder
A. incana	grey alder
Betula nigra	river alder
B. pubescens	birch
B. nigra	river birch
B. verrusoca	silver birch
Elaeagnus umbellata	autumn olive
Larix decidua	European larch
L. leptolepis	Japanese larch
Liquidambar styracifura	sweet gum
Liriodendron tulipifera	yellow polar
Populus deltoides	cottonwood
P. grandidentata	bigtooth aspen
P. tremuloides	quaking aspen
P. x *canadensis*	black Italian poplar hybrids
Pinus banksiana	jack pine
P. contorta	lodgepole pine
P. echinata	short leaf pine
P. nigra var. *Austriana*	Austrian pine
P. nigra var. *calabrica*	Corsican pine
P. resinosa	red pine
P. rigida	pitch pine
P. Strobus	white pine
P. sylvestris	Scots pine
P. virginiana	Virginia pine
Platanus occidentalis	sycamore
Quercus borealis	red oak
Q. rubra	northern red oak
Robinia fertilis	bristly locust
R. pseudoacacia	black locust
Salix x *purpurea*	willow hybrids
Shrubs	
Amorpha fruticosa	false indigo
Eleagnus angustifolium	Russian olive
E. umbellata	autumn olive
Rhux capillina	dwarf sumac

Table 5b. Trees and shrubs known to tolerate alkaline spoil material (Richardson 1977)

Shrubs	
Robinia fertilis	bristly locus
R. hispida	rose acacia
Sarothamnus scoparius	broom
Ulex europaeus	gorse
Trees	
Alnus glutionosa	European black alder
A. incana	grey alder
Betula verrucosa	silver birch

Table 5b. (continued)

Fraxinus excelsior	ash
Juniper	juniper
Larix decidua	European larch
L. leptolepis	Japanese larch
Picea sitchensis	sitka spruce
Pinus banksiana	jack pine
P. nigra var. *Austriana*	Austrian pine
P. resinosa	red pine
P. sylvestris	Scots pine
P. virginiana	Virginia pine
Populus alba	white poplar
P. x *canadensis*	black Italian poplar hybrids
Prunus avium	wild cherry
Robinia pseudoacacia	black locust
Salix alba var. *vittellina*	willow
Salix alba var. *britzensis*	willow
Shrubs	
Atriplex halimus	orache
Colutea aborescens	black senna
Cratoegus monogyna	hawthorn
Eleagnus angustifolium	Russian olive
E. umbellata	autumn olive
Hippophae rhamnoides	sea buckthorn
Lespedeza bicolor	shrub lespedeza
Lonicera maackii	amur honeysuckle
L. tatarica	Tartarian honeysuckle
Ribes aureum	golden currant
· *Robinia fertilis*	bristly locus
· *R. hispida*	rose acacia
Symphoricarpos orbiculatus	coral berry
· *Ulex europaeus*	gorse
Viburnum dentatum	arrowwood

Typical tree density for diverse mixed forest is 2250 trees per hectare (2 m spacing) and 1600 trees per hectare (2.5 m spacing) for tree farms (Powell 1988). Higher density may be used to maximize the production, and encourage the retention, of leaf litter (Celtic Energy Ltd. 1997)

A difficult issue when planning a woodland restoration scheme, is whether or not to grow herbaceous ground cover. Vegetation between the trees may provide desirable supplementary nitrogen. In particular, ground vegetation may be used to stabilize toxic sites prior to forestation. However, an established ground cover will compete with trees for essential nutrients, and thus may be detrimental to growth.

Several approaches may be taken to prevent or remove unwanted ground cover. Once the height of tree seedlings exceeds the height of the ground cover, then cutting or mowing may be effective provided the ground surface is not too rough or steep. To restrict the initial growth of ground cover a plow may be used to turn over vegetation.

Alternatively, herbicides may be applied.

Vogel (1973) notes that there is evidence to suggest that some grasses and legumes are more compatible with trees and shrubs than others. This depends upon rooting characteristics, moisture demands and the growing season.

Fertilizers, lime, mulches and other organic materials may be applied to improve plant growth during early critical years. Fertilizers should be supplied sparingly as excess amounts will clearly encourage vegetative ground cover. Liming is only required for extremely acidic soils. Mulch typically consists of hay, straw and waste wood products reprocessed to a uniform size. Spread on the ground, it encourages the growth of seedlings by reducing surface temperature, controlling soil moisture losses and protecting from prevailing winds. Organic materials may be added to the soil prior to planting in order to provide nutrients and also improve physical characteristics. An example additant is "Biogram", a highly processed sludge, which has been successfully applied at many reforested sites in Wales (Celtic Energy Ltd. 1997). One of the benefits of Biogram is that the most soluble nitrogen forms are rinsed from it during processing. The remaining nitrogen is released slowly over time, rather than being rapidly washed away as leachate.

4 Agriculture

Agricultural land can be broadly classified as cropland, pasture (and hayland) and rangeland. Cropland contains annual, biennial or perennial crops produced on a short-term rotational basis (Powell 1988). Pastureland is used to produce domesticated forage plants for grazing by livestock (American Society for Surface Mining and Reclamation 1983). When grasses and legumes are cut and dried for fodder on a frequent basis, then such land may be classified as hay cropland rather than pasture (Grandt 1988). Rangeland, as commonly seen in the western United States, consists of permanent herbaceous and/or woody vegetation primarily used to provide food for domestic livestock, or indigenous wildlife, without constant management (Powell 1988).

4.1 Pasture

Pastureland is generally the most easy to establish, following soil reclamation, and it is particularly advantageous when soil depths and slopes are such that a continuous vegetative cover is required to prevent erosion (Ries and Stout 1988). In principle a single species or simple mix of species can be planted with minimum effort and expense to provide land stabilization, erosion control and productive pasture.

Powell (1988) outlines the procedure for establishing pasture on reclaimed land. Following contouring, the soil surface may be worked to establish water impounds for the safe capture of excess surface drainage. Topsoil, if available, should be replaced, although reasonable yields of grasses can be achieved by directly planting on non-toxic overburden material.

Following chemical treatment of the soil, a proper seedbed should be established to increase the degree of seed germination and survival. However, there must be some tradeoff between seedbed preparation and site stabilization. Classically, a firm seedbed is desirable. Nevertheless, re-graded or topsoiled mined lands usually have poor soil structure and low infiltration potential. So tillage treatments, e.g., disk or chisel plough, producing increased surface roughness, increased water infiltration and a less dense rooting zone, often give rise to a better long-term yield, even though they do not promote a firmer seedbed.

The species to be planted should ideally seed vigorously, be long-lived, adapt to a range of soil and microclimate conditions, resist disease and insect damage, respond to fertilization and be reasonably drought resistant. Usually a mix of species is required to achieve all of these characteristics. If, as in humid regions, a slow growing grass such as bluegrass is chosen as the dominant species, then faster growing perennials or annual grains may be used in the initial mix to provide erosion control during early months. In semi-arid regions a mixture of cool and warm season species is desirable. The most useful species for pastureland in the semi-arid and arid western USA are crested wheatgrass (*Agropyron cristatum* [L.] Gaertn., *A. desertorum* [Fisch] Scult., and related taxa), tall wheatgrass (*A. elongatum* Host.), intermediate/pubescent wheatgrass (*A. intermedium* [Host] Beauv.), western wheatgrass (*A. smithii* Rydb.), smooth bromegrass and alfalfa (Thornburg 1982).

The seeding time can be very important. It is firstly determined by the species, but also may depend upon microclimate, seedbed preparation, seeding method and the application of mulch. Surface mulches are commonly used to conserve soil moisture and lessen erosion, especially on smooth surfaces.

4.2 Cropland

Grandt (1988) summarizes nine criteria used by the U.S. Soil Conservation service to describe prime cropland:

1. There must exist moisture regimes and available water capacity within a depth of 1 m or in the root zone of less than 1 m, to produce the commonly grown cultivated crops adapted to the region in 7 or more years out of 10 (some regimes are permitted irrigation).

2. Temperature regimes have a mean annual temperature higher than 32°F at a soil depth of 20 in.; mean summer temperature higher than 47°F at a 20 in. depth, with a 0 horizon present; and mean summer temperature higher than 59°F at a 20-in. depth, if no 0 horizon.
3. Soils have pH between 4.5 and 8.4 in all horizons to 1-m depth.
4. There is no water table or the water table is maintained at sufficient depth to allow common cultivated plants to grow.
5. Conductivity of saturation extract is less than 4 mhos/cm and exchangeable sodium percentage (ESP) is less than 15 within 1 m or less of the surface.
6. Soils are not flooded during growing season or are flooded less than once in 2 years.
7. Erodibility factor (K) x percent slope is less than 2 and soil's erodibility factor (I) x climatic factor (C) is less than 60.
8. Permeability rate is at least 0.06 in./hr in the upper 20 in., and mean annual soil temperature at that depth is less than 59°F.
9. There is less than 10 % rock fragments in upper 6 in. coarser than 3 in.

While reclamation of surface-mined lands to these guidelines would be desirable, Powell (1988) considers that such a goal is not completely necessary. It is argued that successful crop growth on reclaimed land can generally be attributed to the existence or creation of physical characteristics that give rise to maximum water infiltration and maximum expansion of the root system. Roots of most agricultural crops can reach depths of 6 ft and achieve horizontal spreading of 4 ft. within a month of growing in a proper environment.

Legumes are perhaps the primary plant species to be used for restoring mined lands to productive cropland use (Grandt 1988). They aid in improving soil tilth, increasing soil nitrogen and improving the numbers, kind and activities of microorganisms, which are important in the decay of organic matter. Alfalfa, sweet clover and birdsfoot trefoil have taproots or modified taproots that adapt well to growth on mined lands.

Forage grasses such as tall fescue (*Festuca arundinacea*), bromegrass (*Bromus inermis*), orchardgrass (*Dactylis glomerata*) and bermuda grass generally establish more slowly on mined lands than do legumes. However, forage grasses are important for stabilizing against erosion and improving soil structure due their fibrous root system.

Small grain crops, e.g. wheat, rye, oats and barley, can be seeded immediately following the re-grading of mined lands. Thus, they are ideal for initial erosion control and as nurse crops for spring-seeded legumes and grasses.

Wheat adapts particularly well to growth on mined soils due to its shallow root system. Grandt (1988) recommends that fungicide-treated wheat should be planted in 8-in. spacings, approximately 1 to 2 in. deep at the rate of 90 lb./acre. It requires large amounts of phosphorus applied in the fall, with 1.5 times the application prior to seeding. Nitrogen should be applied in two application: 20 to 30 lb. N/acre in the fall (drill-applied) and around 50 lb. N/acre in the spring (top-dressed).

High yields of rowcrops have been produced on mined and graded soils even without topsoil. Grandt (1988) reports that yields of corn grown in Knox county, Illinois, USA, eventually exceeded county averages following crop rotation with alfalfa. Corn will perform well under favourable moisture conditions, but is sensitive to moisture stress (Dunker et al. 1982).

Other crops that have been successfully planted on surface mined lands include fruit tress (Seastrom 1965; Fantisch 1973) and vegetables (Mays and Bengston 1978; Jones et al. 1979; Morse and O'Dell 1983).

4.3 Rangeland

Powell (1988) notes that most non-mined native rangelands are not as productive for livestock as they potentially could be. Hence, reclaimed rangeland following surface mining can often be of greater utility. The reclaimed vegetation should, nevertheless, still be similar to pre-mining or nearby vegetation.

Establishment of grasses and forbs is key to productive rangeland. These firstly supply the majority of high quality feed for livestock, and secondly protect the soil from erosion by wind and rain. Suitable grass species must be drought resistant, winter hardy and must fit into livestock and wildlife needs (Asay 1979). Wheatgrasses, needle grasses brome grasses and wildryes have proven suitable in the Northern Great Plains of the USA (Powell 1988). Rangeland in the Southern Great Plains and Intermountain region is predominantly Russian wildrye, mountain bromegrass and wheatgrass. Species in arid areas of the southwestern USA include alkali sacaton, wheatgrasses, Indian ricegrass and wildryes. Blue gamma grass and non-grass species such as alfalfa, sanfoin and yellow sweet clover are also found in all of the above regions.

Supply and control of water is clearly important for establishing ground cover for rangelands in semi-arid and arid regions. Tillage tools can be used to leave very rough surfaces with microridges and small depressions that trap water and give it time to infiltrate. Mulching can also be used effectively to control soil moisture. Supplemental irrigation during the first two-years after planting can promote a long-lasting improvement in species diversity. Typically, irrigation will simulate an above average precipitation year, such as would occur every 2 years out of 10 (Powell 1988).

The establishment of trees and shrubs on reclaimed rangeland adds diversity to the vegetation and generally enhances the utility of the land for wildlife. Some trees and shrubs may naturally invade reclaimed areas, while others may require direct seeding or planting, sometimes with irrigation. Shrubs can be established in arid regions receiving as little as 8 in of rain per year (Aldon

1975; Hassell 1982). Boles (1983) proposes that 1 shrub per 9 m^2 is a sufficient density for rangeland used by both domestic livestock and indigeeous wildlife. For sites developed exclusively for wildlife habitat, a density of 1 shrub per m^2 is recommended in patches covering 5 to 20 % of the total area.

5 Wildlife

Creation of wildlife habitats on reclaimed land may come about under two circumstances (Powell 1988):

1. A critical wildlife habitat was destroyed by the mining operation and so the mining company has as an obligation to restore the habitat close to original conditions. Examples of this type include cases where elk breeding grounds, nesting and feeding sites for eagles, or wetlands have been destroyed.
2. The pre-mining area was a general habitat, but not of crucial significance to one or more particular species. Restoration of wildlife should still be strongly encouraged in such cases, although it may not be a legal requirement. Often, it may be possible to improve upon pre-mining conditions. Habitat enhancement might include planting of seed bearing shrubs, critical placement of brush piles, construction of water resources and creation of borders that did not exist under pre-mining conditions.

A further list of wildlife rehabilitation techniques is given in Table 6.

The choice of vegetation is clearly important. This requires an understanding of the habitat and food preferences of the target wildlife species. Examples of the usefulness to wildlife or various trees and shrubs are given in Table 7. A number of studies of the adaptation of wildlife to reconstructed habitats have been undertaken for various species ranging from large animals down to the smallest microbes. Table 8 provides some example cases. Majer (1989) gives a more comprehensive bibliography of fauna studies in reclaimed lands.

Table 6. Examples of typical wildlife rehabilitation techniques and concomitant benefits (Table 8.1 from Viert (1989))

Technique	Principal benefit
Topogranhic manipulations such as creation of undulating and broken topography, undrained surface depressions, and microtopographic features.	Provides diversity of physical habitat, visual and thermal cover, and the necessary conditions for development of varied biotic habitat features.
Leave intact or modify remaining highwall.	Provides opportunities for cliff nesting raptor species.
Establish or enhance impoundments	Provides still-water habitat, and a source of water for wildlife.

Table 6. (continued)

Direct-haul topsoil for reapplication and spread in nonuniform depths	Native seed and soil microbia aid establishment of floral diversity.
Reconstruct stream channels and streamside habitats.	Predisposes development of riparian systems.
Create large boulder and rock piles randomly distributed about the rehabilitated area.	Provides habitat for small mammals, perch sites for raptors, and substitutes for natural rimrock.
Create large, irregularly shaped brush piles.	Provides temporary substitute for lost micro-habitats and shelters.
Establish shrub/tree plantings along lee sides of rock/brush piles or randomly within rehabilitated area.	Takes advantage of additional snowpack moisture and/or acts as a 'live snowfence'.
Establish shelterbelts within large expanses of rehabilitated grasslands.	Provides wind erosion protection, habitat diversity, and travel routes for wildlife.
Design the shape and spatial distribution of revegetated communities to maximize interspersion and 'edge effect'.	Provides a means to maximize community diversity, wildlife diversity, and wildlife populations.
Select plants to be revegetated based on forage and cover values to wildlife.	Forces consideration of these wildlife needs during seed mix determination.
Improve land management practices to preclude deleterious grazing.	Eliminates habitat degradation due to overutilization by livestock.
Add special structures to rehabilitated areas as necessary (e.g. implanted dead trees for snags, artificial nesting structures, etc.)	Provides for certain physical habitat needs of wildlife species not provided by other techniques.

Table 7. Usefulness to wildlife of various trees and shrubs as cover (C), browse (B), herbage or foliage (H), mast (M), fruit (F) or seeds (S) (from Rafaill and Vogel 1978)

Trees	Uses	Shrubs	Uses
drsEastern red cedar	CFB	Indigobush	CBS
Spruces	CSB	Dogwoods	FBC
Pines	CSB	Hawthorns	CFB
Maples	SB	Autumn olive	FBC
European black alder	CSB	Bicolor lespedeza	SHC
Birches	BSC	Amur privet	FC
Chinese chestnut	MB	Japanese honeysuckle	CBF
Flowering dogwood	FBC	Bush honeysuckle	FBC
Russian olive	FC	Bayberry	FCB
Ashes	SB	Sumacs	FBC
Black walnut	M	Rose	CFBS
Sweetgum	SC	Coralberry	FBC
Tulip poplar	SB	Cranberrybush, arrowwood	FBC
Crab apples	FCB		
Sycamore	SB		
Oaks	MBC		
Black locust	SCB		
Sassafras	BFCS		

Table 8. Example studies of the adaptation of wildlife to reconstructed habitats

Source	Species
Murray (1978)	antelope
Van Waggnor (1978)	small mammals
Parmenter et al. (1985)	vertebrates (cliff-nesting birds, waterfowl, amphibians)
Hingsten and Clark (1984)	small mammals
Allaire (1980)	bird species
Armbruster (1983)	migratory bird species
Phillips and Beske (1989)	golden eagles and other raptors
Holl (1996)	moths
Zarger et al. (1987)	aquatic macroinvertibrates, fish, small mammals, birds
Proctor et al. (1983)	fish and wildlife
Brenner and Helms (1991)	macroinvertibrates
Luff and Hutson (1977)	soil fauna
Starks and Shubert (1982)	soil algae
Armstrong and Bragg (1984)	earthworms
Scullion (1994)	earthworms
Harris and Birch (1989)	microbial activity
Williamson and Johnson (1990)	microbial activity

References

Aldon EF (1975) Techniques for establishing native plants on coal mine spoils in New Mexico. In: Boyer JF (ed) 3rd Symp. Surface Mining and Reclamation, vol 1. NCA/BCR Coal Conference and Expo. II, Washington, D.C.

Allaire PN (1980) Bird Species on Mined Lands. Assessment and Utilization in Eastern Kentucky. Final report 1974-81, Kentucky Univ., Lexington. Inst. for Mining and Minerals Research

American Society For Surface Mining and Reclamation (1983) Glossary of Surface Mining and Reclamation Terminology. 2nd ed, Bituminous Coal Research Inc., Monroeville, Pa.

Armbruster JS (1983) Impacts of Coal Surface Mining on 25 Migratory Bird Species of High Federal Interest, Fish and Wildlife Service, Fort Collins, CO. Western Energy and Land Use Team

Armstrong MJ, Bragg NC (1984) Soil physical parameters and earthworm populations associated with opencast coal working and land restoration. Agriculture, Ecosystems and Environment 11(2):131–143

Asay KH (1979) Grasses for revegetation of surface mining areas in the Western United States. In: Boyer JD (ed) Surface Coal Mining and Reclamation Symp. NCA/BCR Coal Conf. and Expo. V, Washington D.C.

Ashby WC, Vogel WG, Kolar CA, Philo GR (1984) Productivity of stony soils on strip mines, in Erosion and Productivity of Soils Containing Rock Fragments. In: Kral DM (ed) Spec. Publ. 13, Soil Science Society of America. Madison, Wis.

Barnhisel RI (1988) Correction of physical limitations to reclamation. In: Hossner LR (ed) Reclamation of Surface Mined Lands, vol I. CRC Press Inc.

Bertrand AR (1965) Rate of water intake in the field. In: Black CA (ed) Methods of Soil Analysis, vol I. American Society of Agronomy, Madison, Wis.

Boles P (1983) Shrub cover and density on Western rangelands in relation to reclamation success standards for surface mined lands in Wyoming. In: Graves DH (ed) Proc. 1983 Symp. Surface Mining Hydrology, Sedimentology, and Reclamation. OES Publications, Univ. of Kentucky, Lexington

Brenner FJ, Helm J (1991) Macroinvertebrate recolonization and water quality characteristics of a reconstructed stream after surface coal mining in northwestern Pennsylvania, USA. Int J of Surface Mining and Reclamation 5(1):11–15

Byrnes WR, McFee WW, Stockton JG (1980) Properties and Plant Growth Potential of Mineland Overburden. U.S. Environmental Protection Agency, Environmental Research Laboratory, Cincinatti, Ohio

Celtic Energy Ltd. (1997) Draft copy of Bryn Henllys Opencast Coal Site Detailed Restoration and Aftercare Scheme, Aberaman, Wales, UK

Davidson WH (1977) Performance of ponderosa pine on bituminous mine spoils in Pennsylvania. U.S. For. Serv. Exp. Stn. No. NE-358

Dingman SL (1994) Physical Hydrology. Prentice-Hall, Englewood Cliffs, New Jersey

Dollhopf DJ, Postle RC (1988) Physical parameters that influence successful minesoil reclamation. In: Hossner LR (ed) Reclamation of Surface Mined Lands, vol I. CRC Press Inc.

Dollhopf, DJ, Goering JD (1982) Mined land erosion control with surface manipulation. In: Graves D (ed) Symp. Surface Mining Hydrology, Sedimentation and Reclamation. OES Publications, Univ. of Kentucky, Lexington

Dunker RE, Jansen IJ, Thorne MD (1982) Corn response to irrigation on surface mined land in western Illinois. Agric J:74, 411

Fantisch J (1973) Reclamation of areas damaged by mining activities in Czechoslovakia. Ecology and Reclamation of Devastated Lands, vol II, Gordon and Breach, New York

Grandt AF (1988). Productivity of Reclaimed Lands – Cropland. In: Hossner LR (ed) Reclamation of Surface Mined Lands, vol I. CRC Press Inc.

Hadley RF (1961) Some Effects of Microclimate on Slope Morphology and Drainage Basin Development, Geological Survey Research, Chapter B. Professional Paper No. 424-B. Washington, D.C., U.S. Geological Survey

Harris JA, Birch P (1989) Soil microbial activity in opencast coal mine restorations. Soil-Use-and-Management 5(4):155–160

Harwood GD, Thames JL (1988) Design and planning considerations in surface-mined land shaping. In: Hossner LR (ed) Reclamation of Surface Mined Lands, vol I. CRC Press Inc.

Hassell WG (1982) New plant materials for reclamation. In: Aldon EF, Oaks WR (eds) Reclamation of Mined Lands in Southwest, October 20th to 22nd, Albuquerque, N.M.

Heide G (1973) Pedological investigations in the Rhine Brown-Coal area In: Hutnik RJ, Davis G (eds) Ecology and Reclamation of Devastated Land, vol II, Gordon and Breach, New York

Hingsten TM, Clark WR (1984) Small mammal recolonization of reclaimed coal surface-mined land in Wyoming. J Wildlife Management 48(4):1255–1261

Holl, K.D. (1996) The effect of coal surface mine reclamation on diurnal lepidopteran conservation, J. of Applied Ecology, 33(2).

Hossner LR (1988) Reclamation of Surface Mined Lands, 2 vols. CRC Press Inc.

Jansen IJ, Melsted SW (1988) Land shaping and soil construction. In: Hossner LR (ed) Reclamation of Surface Mined Lands, vol I. CRC Press Inc.

Jones JN, Armiger WH, Bennett AL (1979) Specialty Crops – an alternative land use on surface mine soil. Hill Lands – Proc. Int. Symp., West Virginia Univ., Book Office Pub., Morgantown, p 560

Kendle T, Schofield J (1992) Saving our soil. Landscape Design, May issue
Luff ML, Hutson BR (1977) Soil fauna populations. In: Hackett B (ed) Landscape Reclamation Practice. IPC Business Press Ltd., Guilford, England
Lutz JF (1952) Mechanical impedance and plant growth. In: Shaw BT (ed) Soil Physical condition and Plant Growth. Academic Press, New York
Majer JD (ed) (1989) Animals in Primary Succession: The Role of Fauna in Reclaimed Lands. Cambridge University Press
Mays DA, Bengston CW (1978) In: Schaller FW, Sutton P (eds) Reclamation of Drastically Disturbed Lands. ASA, CSSA, SSSA, Madison, Wis.
Medvick C (1980) Tree planting experiences in eastern interior coal provinces. Proc. Trees for Reclamation, U.S. For. Serv. Gen. Tech. Rep., NE-61
Morse R, O'Dell C (1983) Utilization of minesoils for production of vegetable crops. Symp. Surface mining Hydrology, Sedimentology and Reclamation, UKY-BU 133, Univ. of Kentucky, Lexington
Murray FX (ed) (1978) Where Do We Agree? vol II, Report of the National Coal Policy Project, Westview Press, Boulder, Colorado
Nicholson DT (1995) The visual impact of quarrying, Quarry Management, July issue
Parmenter RR, MacMahon JA, Waaland ME, Stuebe MM, Landres P, Crisafulli CM (1985) Reclamation of surface coal mines in western Wyoming for wildlife habitat: a preliminary analysis. Reclamation and Revegetation Res 4(2):93–115
Phillips RL, Beske AE (1989) Distribution and abundance of golden eagles and other raptors in Campbell and Converse Counties, Wyoming. Fish and Wildlife Technical Report, US Fish and Wildlife Service, pp 27, 31
Plass, WT, Powell JL (1988) Trees and shrubs. In: Hossner LR (ed) Reclamation of Surface Mined Lands, vol I. CRC Press Inc.
Powell JL (1988) Revegetation options. In: Hossner LR (ed) Reclamation of Surface Mined Lands, vol II. CRC Press Inc.
Power JF (1978) Reclamation research on strip-mined lands in dry regions. In: Schaller FW, Sutton P (eds) Reclamation of Drastically Disturbed Lands. American Society of Agronomy, Madison, Wis.
Proctor BR, Thompson RW, Bunin JE, Fucik KW, Tamm GR (1983) Practices for Protecting and Enhancing Fish and Wildlife on Coal Surface-Mined Land in the Powder River-Fort Union Region. Science Applications, Inc., Oak Ridge, TN., Bureau of Mines, Washington, DC., Fish and Wildlife Service, Washington, DC. Office of Biological Services Union Region, Mar 83. 261 p
Rafaill BL, Vogel WG (1978) A guide to re-vegetating surface-mined lands for wildlife in Eastern Kentucky and West Virginia. FWS/OWB – 78/74 Fish and Wildlife Service, U.S. Dept. of the Interior, Washington, D.C.
Richards LA (1965) Physical condition of water in soil. In: Black CA (ed) Methods of Soil Analysis, vol I. American Society of Agronomy, Madison, Wis.
Richardson JA (1977) High-performance plant species in reclamation. In: Hackett B (ed) Landscape Reclamation Practice. IPC Business Press Ltd., Guilford, England
Ries RE, Stout WL (1988) Improved Pasture. In: Hossner LR (ed) Reclamation of Surface Mined Lands, vol II. CRC Press Inc.
Russell MJ (1996) Computer program aids design of post-mine topography. Coal-Int 244(2):69–70
Schaefer M, Elifrits D, Barr DJ (1980) Sculpturing reclaimed land to decrease erosion. In: Graves D (ed) Symp. Surface Mining Hydrology, Sedimentology and Reclamation. OES Publications, Univ. of Kentucky, Lexington

Scullion J (1994) Earthworms and soil rehabilitation after opencast mining for coal. Proceedings of the 2nd Int. Symp. on Env. Biotechnol. Brighton, UK, Institution of Chemical Engineers Symposium Series, Publ. by Inst. of Chem. Engineers, Rugby, England, pp 49–51

Seastrom PN (1965) New Land Orchards. Coal Mine Spoil Reclamation Symp., Pennsylvania State Univ., Univ. Park

Starks TL, Shubert LE (1982) Ecological Studies on the Revegetation Process of Surface Coal Mined Areas in North Dakota. 8. Soil Algae. Final report Aug 75-Jun 82, North Dakota Univ., Grand Forks. Project Reclamation, Bureau of Mines, Washington, DC., Jun 82, 75 p

Thornburg AA (1982) Plant Material for Use on Surface-mined lands in Arid and semi-arid regions, SCS-TP-157, U.S. Dept. of Agriculture, Washington, D.C.

United States National Research Council (US. NRC) (1981) Surface Mining: Soil, Coal and Society. Report prepared by the Committee on Soil as a Resource in Relation to Surface Mining for Coal, Board on Mineral and Energy Resources, commission on Natural Resources, National Academy Press, Washington, D.C.

United States Environmental Protection Agency (US. EPA) (1973) Processes, Procedures and Methods to Control Pollution from Mining Activities, EPA-430/9-73-001, US Government Printing Office, Washington, D.C.

United States Soil Conservation Service (1951) Soil Survey Manual, U.S. Dept. of Agriculture Handbook 18

Van Waggonor K (1978) The effects of lignite mining and reclamation on small mammal populations in Texas. In: Samuels DE, Stauffer JR, Hocutt CH, Mason WT Jr (eds) Surface Mining and Fish/Wildlife Needs in the Eastern United States, FWS/OBS-78/81 U.S. Fish and Wildlife Service, Washington D.C.

Ventura JD, Dougherty MT (1980) Design of diversions on surface mine operations In: Graves D (ed) Symp. Surface Mining Hydrology, Sedimentology and Reclamation. OES Publications, Univ. of Kentucky, Lexington

Verma TR, Thames JL (1978) Grading and shaping for erosion control and vegetation establishment in dry regions. In: Schaller FW, Sutton P (eds) Reclamation of Drastically Disturbed Lands. American Society of Agronomy, Madison, Wis.

Viert SR (1989) Design of reclamation to encourage fauna. In: Majer JD (ed) Animals in Primary Succession: The Role of Fauna in Reclaimed Lands. Cambridge University Press

Vogel WG (1973) The effect of herbaceous vegetation on survival and growth of trees planted on coal-mine spoil. Proc. Res. and Applied Tech. Symp. Mined Land Reclamation, Bituminouis Coal Res., Monroeville, Penna.

Williamson JC, Johnson DB (1990) Determination of the activity of soil microbial populations in stored and restored soils at opencast coal sites. Soil Biol and Biochem 22(5):671–675

Zarger TG, Scanlon DH, Nicholson CP, Brown SR, Starnes LB (1987) Ecological Recovery After Reclamation of Toxic Spoils Left by Coal Surface Mining. Phase 2. An Assessment of Environmental Changes Following Intensive Remedial Treatments. Rept. for Jul 75 – Sep 81. Tennessee Valley Authority, Norris. Div. of Land and Economic Resources, Environmental Protection Agency, Cincinnati, OH. Hazardous Waste Engineering Research Lab., Jul 87, 144 p

Mining Lakes: Generation, Loading and Water Quality Control

Helmut Klapper

c/o UFZ Centre for Environmental Research Leipzig-Halle, Dept. of Inland Water Research, Brückstr. 3 a, D-39114 Magdeburg, Germany

1 Hydrography and Geology in Relation to Water Quality

During the surface mining process the overburden layers tend to become mixed and former aquifers destroyed or at least heavily changed. With current technology, the quaternary and tertiary layers above the coal seam are cut in by bucket wheel excavators and transported by a conveyor belt bridge into the empty part of the mine. Other excavators take off the lignite layer for transportation out of the mining hole. The volume of a potential mining lake resulting from this process is the deficit from the removal of coal. Sometimes the volume is smaller, because the ash of the coal, burnt in nearby thermal power plants, or overburden of other newly opened mining sites is dumped in the mining hole.

Today, the opencast process starts with dewatering the whole mining area down to the layers under the coal to be mined. By a series of wells and well batteries the water table is lowered in order to dry out the lignite so as to improve its burning value. Depending on the local conditions the ratio of withdrawn water to extracted lignite may be from 100 % to more than 1000 %.

The hydrologic conditions in rivers and drainage basins are heavily disturbed from the early beginning of preparation of an opencast mine. Larger rivers crossing the mine site have to be transferred into new canal beds surrounding the site and, being unnatural, tightened to avoid seepage of water into the dewatered underground. Smaller rivers and creeks and wetlands become dry. In connection with the geogenical acidification of future mining lakes, the change of the redox-potential in the underground is of special interest. Upon dewatering, the sulfur-containing minerals pyrite and marcasite come into contact with air and therefore with oxygen, instead of oxygen-poor groundwater. Before and during the mining operation, the virtually continuous addition of groundwater into nearby rivers causes unnatural balanced throughflow. No typical low water situations occur for the decades of mining activities. In addition, the water temperature is more balanced by the groundwater influx. Within the long time span of a mining operation, the water balance of the whole landscape, with its forests and wetlands, as well as the water users and water management, must adapt to the changed hydrology. In Germany, a typical case is the mining region Lusatia and river Spree. Mining operations have affected a remarkable and unique wetland called Spreewald and downstream Berlin's largest drinking water treatment plant at lake Müggelsee, fed by the river Spree.

Rapid closure of many mines after German unification introduced, within a short time, challenges of satisfying the water demand for filling newly forming lakes and for refilling the groundwater deficit around the open air mining holes. The volume of the emerging 160 mining lakes amounts to 7.5×10^9 m^3 and the deficit in the emptied groundwater layers is 13.5×10^9 m^3. Together the total water demand is 21×10^9 m^3 (Ziegenhardt 1994).

By 1990 about 2500 km^2 of land was impaired by a lowered groundwater table. Now the active mining is concentrated at only a few sites and the decreased groundwater abstraction may lead to some complications for Berlin's water supply and the wetland-ecosystems of the Spreewald. Several of the new mining lakes are, therefore, being developed as reservoirs for water storage, with inlets and outlets constructed. This way, low water enrichment is possible, to avoid or to minimize the negative effects during the time of readjustment to the natural water balance (Ziegenhardt and Trogisch 1996).

Those lakes emerging in the recent, largest open pits will, in particular, exceed many of the desirable hydrological characteristics of the majority of the natural lakes in the Baltic region of northern Germany. The mining technology creates hollows with relatively steep slopes. Therefore, the bioproductivity of the littoral zone in subsequent generated lakes is low. Also, in the first year of existence, the beds of mining lakes typically consist of minerals with a very low content of organic matter. A couple of years is needed until the layers of sand and coal on the bottom are covered by dead algae, forming the typical "gyttja"-mud, well known from natural lakes where it serves many animals as a habitat and food source.

Table 1. Mining holes in eastern Germany to be flooded within the few next years. Group 1 with mean depth > 15 m according to hydrological class 1; group 2: 10 to 15 m according to class 2 of standard TGL 27885/01 (after BLAG 1994)

Mining lake Group1	Volume [x 10^6 m^3]	Area [ha]	Mean depth [m]
Berzdorf	387	950	40.7
Nachterstedt	226.3	630	35.9
Klinger See	102	300	34.0
Greifenhain	330	973	34.0
Spreetal-NO	97	314	30.9
Cospuden	107.6	393	27.4
Espenhain-Störmtal	154.6	560	27.6
Espenhain-Markkleeberg	65	250	26.0

Table 1. (continued)

Gräbendorf	93	412	22.5
Geiseltal	409	1890	21.6
Zwenkau	188	900	20.9
Kleinleipisch	23	124	18.5
Ilsesee-Meuro	134	754	17.8
Restsee Scheibe	133	714	18.6
Gröbern	70.4	400	17.6
Witznitz II-	97.8	560	17.5
Merseburg-Ost	82.5	520	15.9
Wulfersdorf	22	141	15.0
Group 2			
Witznitz II Kahnsdorf	17.7	120	14.8
Bärwalde	14.8	1017	14.6
Holzweißig-West	23.3	160	14.6
Goitsche	217.5	1500	14.5
Kayna Süd	28	200	14.0
Breitenfeld	27.5	200	13.8
Königsaue	30	220	13.6
Skadoer See	130	980	13.3
Bergheider See	38	290	13.1
Rösa	97	750	12.9
Spreetal-Bluno	152	1211	12.6
Dreiweibern	35	286	12.2
Drehnaer See	17	142	12.0
Sedlitzer See	141	1311	10.8
Golpa Nord	66.1	620	10.7
Lichtenauer See	25	234	10.3
Lohsa II	99	958	10.3
Bockwitz	19.0	187	10.2
Werben	9.1	90	10.1
Burghammer	43	432	10.0
Olbersdorf	6	60	10.0

Regarding the utilisation of water for drinking, fisheries or recreation, the lakes of special interest are those which possess, from the beginning and also in future, a sustainable good water quality. One of the most important prerequisites for developing this type of "clear water lake" is a great depth. In Table 1 the newly forming German mining lakes are arranged in order of their mean depth. Those with more than 15 m depth have the potential to be class 1 and those between 10 – 15 m to be class 2 according to the German TGL 27885/01 (1982) standards. In principle, there are two different options for developing mining lakes. From an human view the lake should contain the cleanest water possible, so it is suitable for purposes such as drinking water, bathing, or fisheries (Fig. 1 and Table 2). However, with respect to nature protection and landscape formation, quite different targets will be set. Because of the general agricultural overproduction in the European Union there is no pressure to obtain maximum yield from each hectare of land. Therefore, some of the new mining lakes may be developed only to provide interesting parts in the landscape, with a high diversity of biological species. To meet this target, a very diverse morphometry of the new lake has to be established. The lake should possess deep as well as shallow parts, a long shoreline relative to the lake area, bays, islands, a mixture of steep and shallow banks etc (Fig. 2). In this way many diverse habitats for different plants and animals are available. In the case of a lake with no human use or special water quality demands, a higher bioproductivity is no disadvantage. In other cases the environment might even remain in an acidic state without fish, mollusc or higher crustaceans and, nevertheless, be an interesting object of limnology with respect to conditions in extreme habitats.

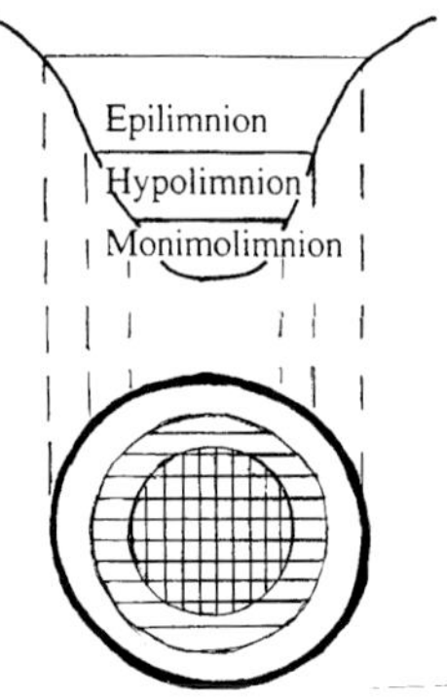

- great depth
- very large ratio of hypolimnion to epilimnion volume
- stable stratification
- steep shore
- small part of the benthos belongs to the epilimnion
- low shore development
- small nutrient import (inflow)
- lake bottom with phosphorus binding matter
- low nutrient recirculation

Water suitable for very demanding water uses: Raw water for drinking and process water, bathing, diving, coregonid fishery

Fig. 1. Hydrography and quality factors for low productive clear water

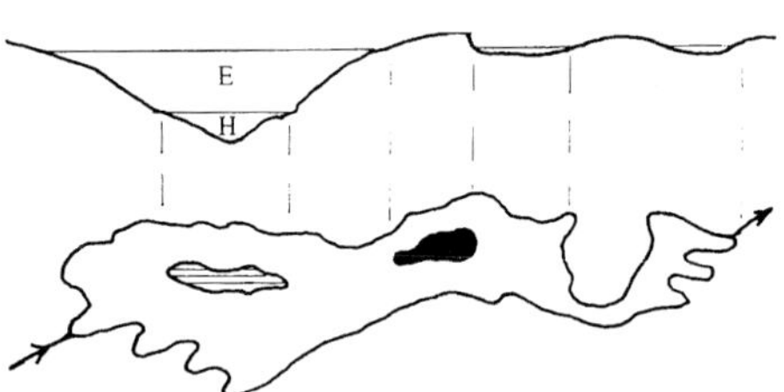

- shallow depth
- small ratio of hypolimnion to epilimnion volume
- lake bottom is mainly in the epilimnion
- very shallow banks
- considerable shore development
- diverse morphology
- high nutrient inputs
- high nutrient recirculation

Water unsuitable for demanding uses. Water body suitable for nature protection and as parts of the landscape worth seeing

Fig. 2. Hydrography and quality factors for high bioproductivity and high species diversity

Table 2. Characteristics of a clear-water lake with low bioproductivity

Hydrography	Chemistry	Biology
large mean depth	poor in macronutrients: P, N, Si	low primary productivity
large maximum depth	precipitation of P with binding cations Fe^{3+}, Al^{3+}, Ca^{2+}	limitation of indispensable resources: light, P, N, Si (at pH < 3 also DIC and DOC)
large lake area	matter flux directed to the sediment	at pH < 3 all species that need carbonate and hydrogencarbonate are excluded
large lake volume	low import of matter	phytoplankton is curbed by
Small drainage basin	matter export high	• sinking losses in macrophyte standings • grazing losses • biogenic calcite precipitation • flushing losses
few inhabitants, or equivalents, in the drainage basin	binding to permanent sediment greater than redissolution	
extensive utilization of the drainage basin	hypolimnion aerobic	
steep shore slopes	stable fixation of phosphorus in oxic sediments	
small part of the bottom belongs to the littoral zone	with pH < 3 no HCO_3^- and CO_3^{2-}	bioflocculation of bacteria and colloids
minimal bank development	toxic influence of acid-soluble heavy metals (?)	low biodiversity
stable stratification		
water exchange only with groundwater		

2 Water Origin: Quantity and Quality of Ground and Surface Waters

Tables 3 and 4 list the advantages and disadvantages of filling mining holes with ground and surface waters. In regions where the lignite and overburden have a low sulfur content and possess enough carbonate hardness for neutralisation of occasional geogenically caused acidification, groundwater is best suited as the filling water for generating a lake. Use of groundwater guarantees a lake with sustainable good water quality, i.e. low bioproductivity and high resistance against eutrophication and its impairing consequences (Table 3). Some of the disadvantages of groundwater filling may be avoided. Instead of flooding with the very slowly rising underground water at the site, additional groundwater from the dewatering of active mines, not too far distant, may be used. However, this foreign water has to be analyzed and its suitability confirmed. Excellent groundwater may be obtained from the dewatering before mining. The wells in the undisturbed rocks in general contain circumneutral waters, sufficiently buffered by carbonate hardness. The phosphorus and heavy metal contents are low. Rapid flooding with such neutral groundwater, so that the water table in the lake rises faster than in the surrounding underground, seems to be the most beneficial solution.

This practice has been utilized in the lignite mining region south of Leipzig, Germany. Instead of filling with river water from the Weisse Elster, the quality of which would have been suitable only after a treatment, a 23.5 km long pipeline has been built. Currently, the pipeline delivers a flow of 45 m^3/min, from the active mine Profen, to flood the former mine Cospuden. The pipe system will be extended to 70 km and completed with side branches in the coming years to fill the mining holes of Markkleeberg, Haubitz, Hain, Störmthal, Werben and Zwenkau (Fig. 3, Zeh 1998).

Table 3. Advantages and disadvantages of filling the mining hole with rising ground water

Advantages	Disadvantages
Provided sufficient bicarbonate, the rising groundwater is poor in phosphorus, which is bound to Fe^{3+}, Al^{3+}, Ca^{2+} and clay minerals	Filling times are extremely long in extended dewatered mining sites and with locally destroyed aquifers
Low organic mother load is responsible for poverty of heterotrophic (and pathogenic) germs	Hydraulic gradient from a higher groundwater table in the surrounding to the lower lake surface endangeres the stability of the slopes
Metal hydroxides are precipitated in the lakes; phosphorus and bacteria are transferred with the precipitates to the bottom sediment	Pyrite is oxidized in the area of groundwater exfiltration, especially in the overburden heaps. Sulfuric acid together with acid-soluble metals impair the water quality
Favourable trophic state for many years, relating to the morphometry	Geogenically acidified waters are not inhabitable for many organisms including fishes
If not impaired by acidification, the water is qualified for particular demanding utilization from the beginning of the filling process	After neutralization, the previously dark-red, dissolved ferrihydroxide of acidic lakes precipitates as ochre coloured turbid flocs. This ugly stage of succession towards more nature-near conditions may last for decades of non-usability

Table 4. Filling the mining hole with river water

Advantages	Disadvantages
Quick flooding possible; fishes survive from the first filling stages	Many running waters are highly loaded with nutrients, oxygen consuming and hazardous substances
From the first flooding neutral and relatively poor mineralized water	Poor trophic conditions exist in the first years, relating to the morphometry.
Hydraulic gradient from the higher lake level to the lower groundwater table stabilizes the slopes	Loading of the surface waters may be introduced into the groundwater space with the danger of irreversible damages

Table 4. (continued)

By oxygen-consuming substrates, the underground environment becomes anoxic and metals are immobilized in sulfidic form; the pH of the groundwater rises	Mass development of algae, because of eutrophication, impairs the recreational water utilization
By incorporation in algae biomass and sedimentation, nutrient and hazardous substances are transferred permanently into the bottom sediments	Oxygen depletion in deep water layers reduces the habitat for fishes and their food animals
Not so demanding utilization (e.g., watersport without body contact with the water) is possible at an early stage	Especially in the first years, technological requirements to improve the water quality by treatment of filling water or by in-lake ecotechnologies

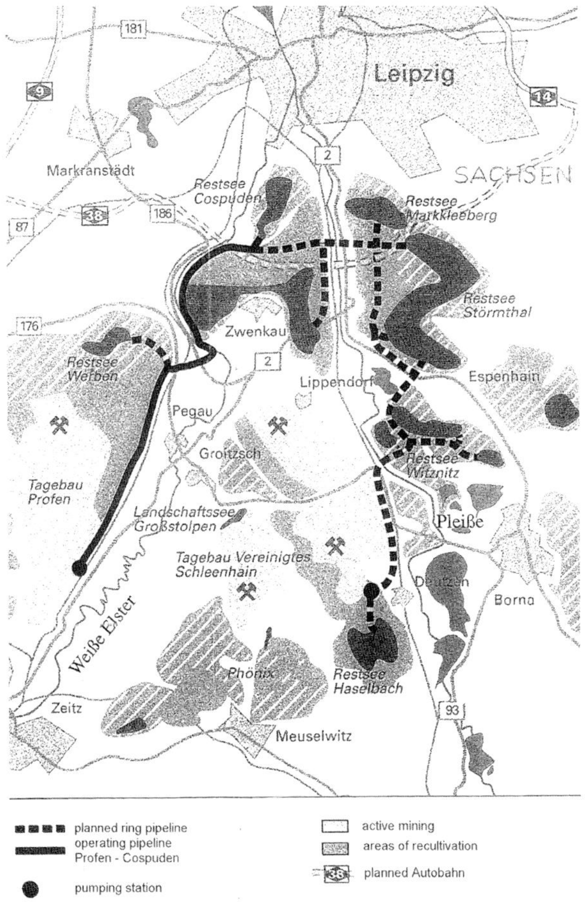

Fig. 3. Concept for flooding mining lakes south of Leipzig, Germany (from Zeh 1998)

The flooding of mines with surface waters may be undertaken either for filling purposes only or alternatively to give a partial or fully permanent throughflow. An example of the latter case is the Mulde Reservoir near Bitterfeld, Germany. The former mining hole Muldenstein now functions as a river water treatment plant. The diverse water quality parameters are eliminated with varying efficiency. This depends on the behaviour of the matter with respect to adsorption, sedimentation, introduction into nutrient chain and metabolization or bioproduction of autochthonous biomass and photosynthetic oxygen (Table 5).

The effect of the Mulde Reservoir in Germany is comparable to that of a pre-basin to a drinking water reservoir, protecting the main basin against the import of unwanted matter (TGL 27885/02 1983). On the eastern bank of the Mulde Reservoir, which consist of loose rocks from old overburden heaps, are many sources of acidic waters with high contents of dissolved iron. Mixed with the neutral, bicarbonate containing reservoir water, ferric hydroxide is precipitated, obviously together with phosphate.

Therefore, the reservoir may be compared not only with a pre-basin, but also with a phosphorus-elimination plant in which phosphorus is separated by sedimentation and deposition on the bottom of the water body. Together with the settling matter, bacteria may also be transferred from the water into the permanent bottom sediments. Consequently, the outlet of the Mulde Reservoir is more appropriate for lake filling water than the Mulde-river upstream. Therefore, since 1998, filling water has been taken from the outlet of the reservoir to flood the various basins of the big mining complex Goitsche near Bitterfeld, within a short time period.

Table 5. Mining lake Muldestausee, Germany utilized as a as river water treatment plant (yearly means from Gewässergütebericht Sachsen-Anhalt 1992)

Criteria	Inlet samples	Arithmetic mean	Outlet samples	Arithmetic mean	Elimination [Inlet = 100 %]
suspended matter [mg/l]	25	17.2	25	2.3	86.6
inorganic nitrogen [mg/l]	25	6.5	24	6.1	6.15
orthophosphate-P [mg/l]	25	0.260	25	0.174	33.1
total phosphorus [mg/l]	22	0.924	22	0.378	59.1
oxygen [mg/l]	25	10.2	26	11.0	7.84
O_2-saturation [%]	25	90.0	26	99.0	10.0
BOD_5 [mg/l]	25	4.9	26	4.1	16.3

Table 5. (continued)

chem. oxygen consumt [mg/l]	19	6.2	17	4.8	22.6
chem.oxygen demand [mg/l]	24	20.4	24	14.8	27.5
total organic carbon [mg/l]	12	7.6	12	6.1	19.7
total zinc [µg/l]	13	136.0	13	85.0	37.5
total copper [µg/l]	13	7.6	13	3.8	50.0
total cadmium [µg/l]	13	2.6	13	1.5	42.3
total iron [µg/l]	13	660.0	13	242.0	63.3
total manganese [µg/l]	13	205.0	13	155.0	24.4

The bottoms of the two northern basins Niemegk and Mühlbeck in the Goitsche mine/Germany consist of pyrite-rich amber sands, containing a high acidification potential. The possibility of acidification of the newly forming lakes in the region was estimated after soil and water monitoring, several elution experiments, computation and modelling. Depending on the depth of the acidity exchange at the bottom, the acidification may be stopped during the first flooding. If this is not the case, some special measures against the acidification must be performed. It should be mentioned, that the buffering capacity of the Mulde river is relatively low, due to low carbonate hardness (Kringel et al. 1997, Kringel 1998).

The following water quality problems have to be considered, according to the origin and kind of the filling water, the shape (especially the depth) of the lake basin, its position within the regional groundwater field and the nature of the geological substratum (Fig. 4):

1. Acidification is the most severe problem in connection with groundwater filling, especially in regions with sulfur-rich lignite, as in the case of the Lusatian mining district of Germany. Due to the oxidation of pyrite and marcasite, not only has low pH been observed, but also a high content of iron and manganese and, if present in the rocks in question, other metals. The report focuses on this problem in Chapter "Loading of Mining Lakes and Water Quality Control. 1 Acid Mining Lakes".
2. Salinization is important when mining is performed in close proximity of tertiary lignite and salt layers. The mining lake Merseburg-East in Germany had deep waters containing three-fold the oceanic concentration of sodium chloride. Currently the concentration is two-fold. Within the deep water, of high specific density, lignite particles are free floating rather than settling to the bottom.
3. Contamination with hazardous substances may occur when the open-air mine had been used as a dumping site for industrial or urban wastes, when

surface water used for filling is polluted or when geogenically acidified ground water dissolves heavy metals on its way through the overburden heaps.

4. Eutrophication is the main problem when the pit becomes filled with nutrient-rich surface waters. Algae mass development and turbid water in a mining lake restrict its use for most purposes. However, as shall be discussed later, in the case of acidification a controlled addition of eutrophying phosphate may shorten the succession from acidic to neutral lakes.

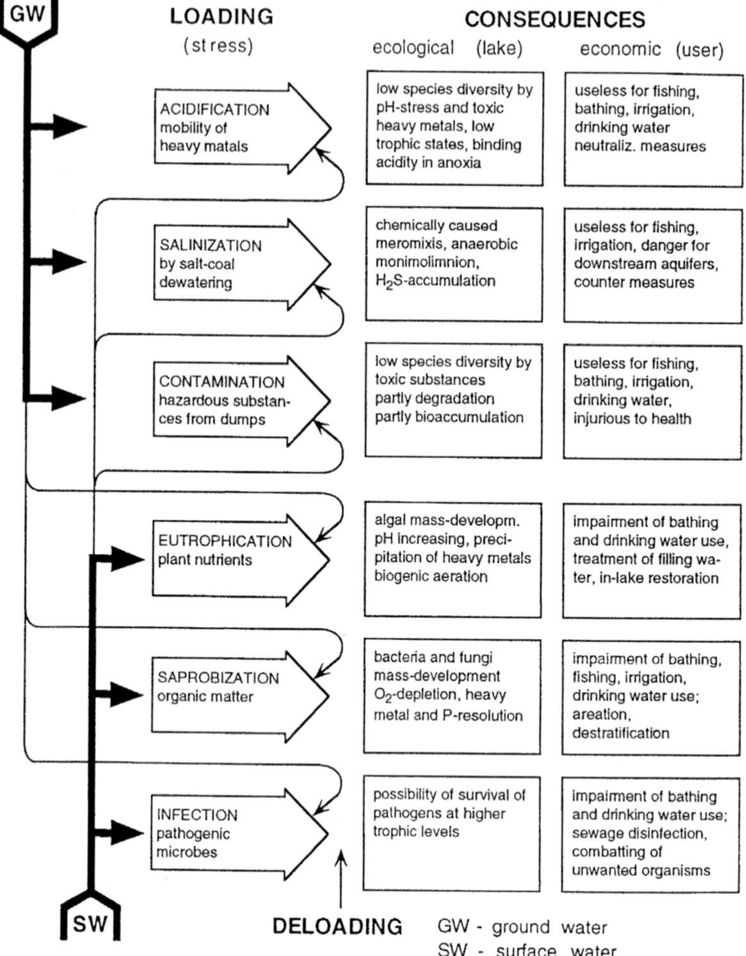

Fig. 4. Types of loading and consequences in mining lakes filled by ground (GW) and surface water (SW) (from Klapper et al. 1996)

5. Saprobization, i.e., loading with allochthonous and autochthonous degradable organic matter, is actually of lesser importance, at least in Germany. The water quality of the rivers used for filling has improved during the last years due to higher standards of sewage treatment. Waste is no longer dumped into mining holes because coal processing industries have been closed.
6. Infection with pathogens is related to the pollution of the filling waters and distinct types of water utilization. Duck feeding on a mining lake precludes bathing, because of *Salmonella* infestations.

Mixing of ground and surface waters allows considerable control of water quality in the lake. To abate eutrophication, a high percentage of groundwater is needed in the filling of the lake. With a content of cations such as Fe^{3+} and Al^{3+}, phosphorus will be bound and transported to the bottom sediments. Similarly, contaminants and bacteria may be adsorbed and precipitated.

In the case of the eutrophic mining lake Kayna-South in Germany, iron-rich groundwater is added and mixed using deep water aerators. Decreasing concentrations of phosphorus have been recorded. The "mesotrophy" water quality target should be achieved upon reaching the final water level.

Conversely, when adding surface waters, salty water may be diluted and acidic waters neutralized by carbonate hardness. The Lusatian lignite district is especially affected by geogenic acidification. Consequently, filling with river water was elected as the main way to cope with the problem.

The large opencast holes around Lauchhammer, Senftenberg, Burghammer and in the Schlabendorf district of Germany are presently being filled with uprising acidic groundwater. Values of pH less than 3 and high base binding capacities call for urgent action. Construction of pipelines is necessary, to allow filling with surface waters. This would achieve a water mixture with diminished acidity, and would stop further acidity influx, by forming a lake level higher than the surrounding groundwater table.

3 Living Conditions in Mining Lakes

3.1 The Chemical Environment

Compared with natural (glacier) lakes, the chemical environments of mining lakes, in the first years of their existence, are more influenced by the geological surroundings, the shape and size of the hole, the origin of the filling water, relatively high salt contents, and especially sulphate and heavy metals.

The mining lake bottoms mainly consist of minerals with low organic carbon, but with high phosphorus binding ability. There is only little growth of bacteria and algae on the bottom and nearly no closed biofilm for biological self-tightening against the surrounding groundwater. In the case of bank filtration for drinking water purposes, particulate matter, such as algae, pass through the biologically inactive sand layers of the shore and appear unwanted in drinking water. The consequences of different types of loading have been discussed earlier. The most important influence on the aquatic environment originates from geogenic sulfur acidification. The main processes are as follows: An initial weathering of iron-disulphide occurs with the first direct contact of pyrite-containing layers with the air (and water):

$$FeS_2 + 3.5\ O_2 + H_2O \Rightarrow Fe^{2+} + 2SO_4^{2-} + 2H^+$$

The Fe^{2+} ions are further oxidized to Fe^{3+} by air-oxygen:

$$Fe^{2+} + 0.25\ O_2 + H^+ \Rightarrow Fe^{3+} + 0.5\ H_2O$$

This reaction is strongly accelerated by microorganisms of the type *Thiobacillus ferrooxidans*. The next step, the formation of ferrihydroxide, takes place mainly at the transition from the heaps to the open water and is connected with the highest share of proton delivery (Fig. 5).

$$Fe^{3+} + 3H_2O \Rightarrow Fe(OH)_3\ (\Downarrow) + 3H^+$$

In the very acidic state, the ferric hydroxide prevails in solution. Therefore, the lakes with pH values less than 3.5 remain with clear water of a dark red-brown colour (Fig. 6).

The overall process of geogenic sulfur acidification may be summarized by the following simplified equation:

$$FeS_2 + 3.75\ O_2 + 3.5\ H_2O \Rightarrow Fe(OH)_3 + 2\ SO_4^{2-} + 4H^+$$

The various methods of neutralization, natural and artificially enhanced, will be discussed later. However, a few remarks may be given about the consequences of raising the pH of the aquatic environment:

- The main change is the decreasing solubility of ferric hydroxide. The water becomes turbid. The colour varies to ochre. The Secchi-depth decreases.
- A co-precipitation occurs, due to which the most heavy metals, bacteria and phosphorus are transferred permanently into bottom sediment.
- At the end of the process the mining lake get its best water quality within the whole succession from the first upcoming groundwater until the final "climax" stage.
- Within the pH-range of the bicarbonate buffer the supply with organic carbon is guaranteed by algae assimilation and, from this time on, phosphorus becomes the limiting factor controlling the primary productivity.

The water chemistry of the mining lakes is governed by three buffering systems: the circumneutral bicarbonate buffer, the acidic aluminium buffer, and the iron buffer. Alkalinization does not change the pH unless the base-binding capacity is saturated.

When this occurs, the pH shifts from one to the next buffering system. The changes are similar to titration curves. In connection with acid rain, the aluminium buffer is dominating. The affected waters prevail in the pH-range 3.6 – 4.2. In mining lakes the iron buffer is absolutely dominant at pH values < 3.8 (Ulrich 1981).

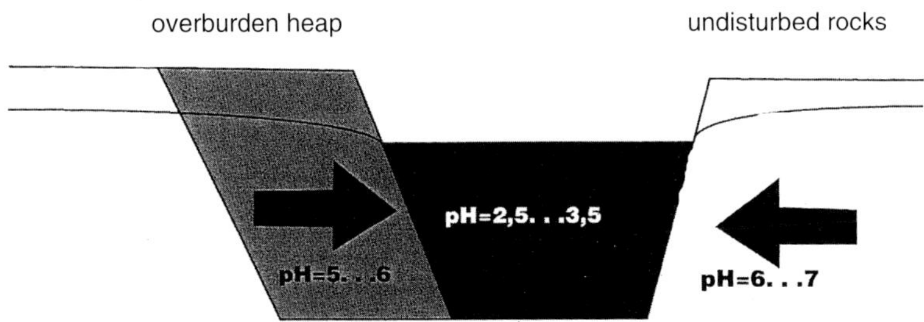

Fig. 5. pH-values in mining lakes and adjacent aquifers (adapted from Reichel and Uhlmann 1995)

Fig. 6. Acidic groundwaters with dissolved ferric hydroxide, filling the new mining lake Skado, Germany

In the Lusatian browncoal region of Germany, most lakes were found to be at about pH 2 to less than 4 , with others between 6 and 8. One exception, the Felix See, was observed to have a pH of about 4. In this case the concentration of aluminium was ten times higher than that of iron. The conditions in this special case resemble those in lakes influenced by acid rain (Fig. 7).

3.2 Habitat for Plants and Animals

From the immense literature available on acid rain, it seems that hardly any life can be expected at a pH less than 4. Indeed, the youngest geogenically acidified mining lakes with a pH of 2 – 3 are abiotic at first glance. Nevertheless, such an extreme habitat has been shown to be colonized by specialists, some in great abundance because of the lack of competition.

Organisms able to survive in the extremely acidic environment possess mechanisms for detoxification by ion exchange, by oxidation and flocculation of heavy metals, by excretion of chelates, and by transformation of metals into other dissolved forms with lower toxicity. Some algae respond with dormancy or concentrate toxicants in terminal cells that they later discard (Lüderitz 1988).

Compared to natural lakes, acid mine lakes generally have low primary productivity. Because of the lack of bicarbonate in acidic lakes, algal groups of the *Scenedesmus* photosynthetic type are absent. The first phytoplankton to appear are from the same groups to be found in bog waters: chloromonads, cryptomonads, dinoflagellates, etc.

Important processes occur at the level of the pico- and bacterioplankton. The total algal biomass is low, and this criterion alone would accord with a classification as "oligotrophic" (Nixdorf et al. 1995). This low algal production is unexpected, considering the relatively high phosphorus supply. Obviously factors other than nutrients limit productivity. Moreover, the seepage waters from overburden heaps with a pH < 2 are not really abiotic. At least bacteria occur in these waters. The pyrite oxidation itself is described as a chemo-autotrophic microbiological process by some species of *Thiobacillus*.

Besides algal development, heterotrophic degradation is strongly inhibited in the acidic environment. Leaves that have fallen from surrounding trees into the acidic lake remain undegraded for years on the lake bottom. Trees submerged during the filling process persist for decades as undegraded and quasi "acid-preserved." The low respiration rates and low solubility of CO_2 in acidic waters seem to be the reasons for carbon limitation of bioproductivity (Ohle 1981, Schindler 1994). Macrophytes are lacking only at a very early filling stage.

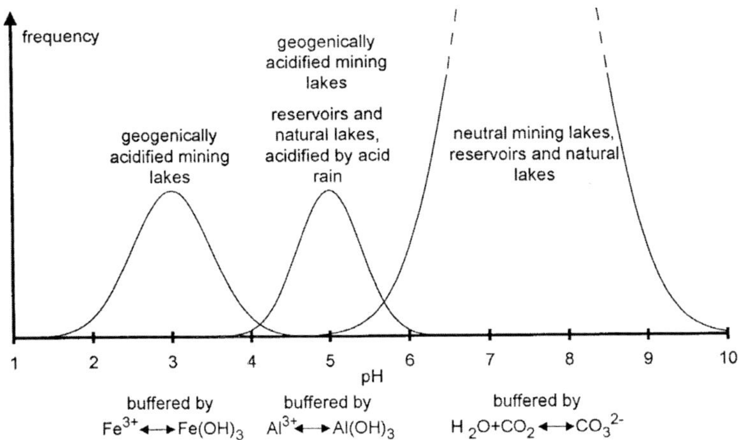

Fig. 7. Generalized frequency distribution of lakes with different acidity in Germany (from Klapper et al. 1996)

Uprising groundwater rich in sulfuric acid, iron and sulphate, a changing shoreline, and an unstable subsurface during the first quick filling phase preclude macrophyte growth for hydrochemical and edaphic reasons. Soon after the hydrological regime has stabilized, however, the typical pioneer plant *Juncus bulbosus* occurs. In this "one species biocenosis" according to biocenotic laws, *J. bulbosus* is very frequent in the littoral zone and also on the lake bottom, if the water is sufficiently clear. Photosynthesis by CO_2-consumption leads to iron flocculation on the stems. Occasionally the ochre precipitation pulls whole plants down to the bottom. This species, which is mainly vegetatively reproducing, sprouts a second time during summer and appears again grass-green in colour. On sunny days the raised pH because of CO_2 assimilation can be seen with naked eyes: in the neighbourhood of the *Juncus* stands iron hydroxide is precipitating (Fig. 8).

Pietsch (1993) distinguished some typical communities characterized by the dominant species besides *Juncus bulbosus* stands: *Potamogeton natans*, *Utricularia minor*, *Sphagnum cuspidatum*, *S. obesum*, *S. inundatum* and the fern *Pilulifera globulifera*. He described a further transition to a stage with neutral water containing bicarbonate and high species diversity. Of the animals, those groups which cannot exist in an acidic environment are those which need calcium carbonate for their skeletons or shells, such as fish, amphibians, snails, mussels and higher crustaceans. During the first filling stages, acid lakes are also without zooplankton. The rotifers (especially *Brachyonus urceolaris*) are inhabiting lakes with $pH < 3$, as does *Chydorus sphaericus*. *Cyclops* was found at pH 3.5 and higher. *Daphnia* occurred only in circumneutral lakes. Whether acidity excludes them or their absence is caused by the high frequency of invertebrate predators like *Corixa* is not clear (Tittel, pers. comm.).

Fig. 8. Mining lake Roter See near Burgkemnitz, Germany. Iron hydroxide is precipitating due to CO_2-consumption around *Juncus bulbosus*

4 Succession of Water Quality

During the primary filling of new mining lakes and in the first years of their existence, the water quality undergoes considerable change. The rising water level causes a succession from a shallow to a deep lake. The first shallow stage shows no stratification and no oxygen gradients from the surface to the bottom. In the next stage, the first stratification may be critical, because the hypolimnion is extremely small in relation to the epilimnion. Degradable material concentrates in a very small volume and may cause oxygen depletion in the deep water (Fig. 9).

In connection with acidified lakes this oxygen depletion may be desirable. However, unfortunately in this case the autochthonous production of organic matter is usually too low to achieve a depletion of oxygen. With further deepening of the lake, the hypolimnion: epilimnion ratio rises. This stage is wanted in neutral lakes, but unwanted in acidic ones, which need places with anoxic conditions, where microbial neutralisation by desulfurication may occur.

The deepest, wind-protected lakes, as well as the chemically stratified lakes, are susceptible to meromixis, i.e., the deepest water layers are excluded from the usual turnover-phases of the dimictic lakes in spring and autumn. In such cases, the unmixed monimolimnion becomes anaerobic, thus, having preconditions for desulfurication at depth. Nevertheless, meromixis alone is not yet an ecotechnology for microbial neutralization. Only the anaerobic monimolimnion turns to a neutral pH. Due to lack of circulation, the surface-near water layers also remain acidic for decades (e.g., lake Waldsee near Döbern, Germany). Typical sequences of the main characteristics of the water body during the filling with surface water are shown in Fig. 10.

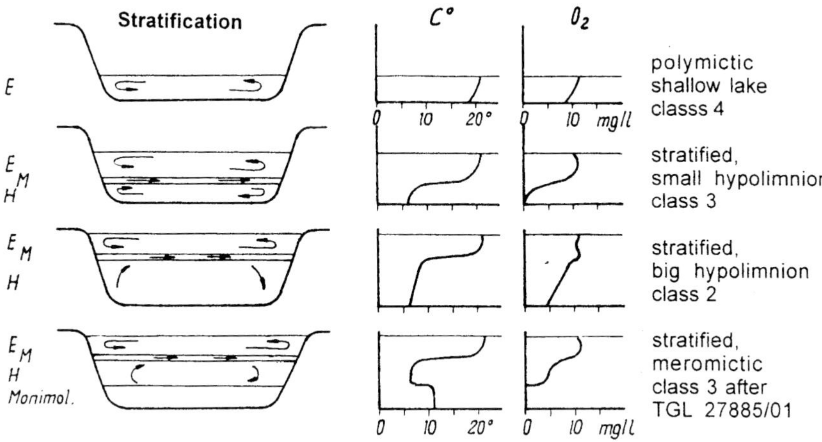

Fig. 9. Change of the stratification along with the filling level of a mining lake (from Klapper 1995)

When mining wells are taken out of operation, a fast rising of groundwater occurs due to the high hydraulic gradient. With the introduction of surface water, the water level in the lake rises faster than the water table of the surrounding groundwater. With the flow-direction from the lake into the underground it is possible to avoid further groundwater entrance and acidification. Towards the completion of filling the lake the introduction of surface water should be finished. Depending on the origin of the groundwater passing through the filled lake, a secondary acidification could then occur. However, there will be no problems with acidic groundwater as long as the acid binding capacity is not used up. If this happens, technologies for neutralization should be considered (see Chapter "Loading of Mining Lakes and Water Quality Control. 1 Acid Mining Lakes").

Similar to natural lakes, the groundwater throughflow is typically accompanied with changes of the water quality because of distinct trapping effects. In the case of lignite mining lakes, the content of, notably, iron, manganese and phosphorus in the entering water decreases due to oxidation and flocculation processes. Organic matter and biomass produced by photosynthesis, nutrients, etc., are incorporated into plankton biomass and transferred to the bottom sediments. The water leaving the mining lake and entering the underground again to become a part of the groundwater, may become anoxic due to the degradation of biologically produced organic matter. The consequences in this case are, for example, denitrification and fixing of sulfidic iron (Fig. 11).

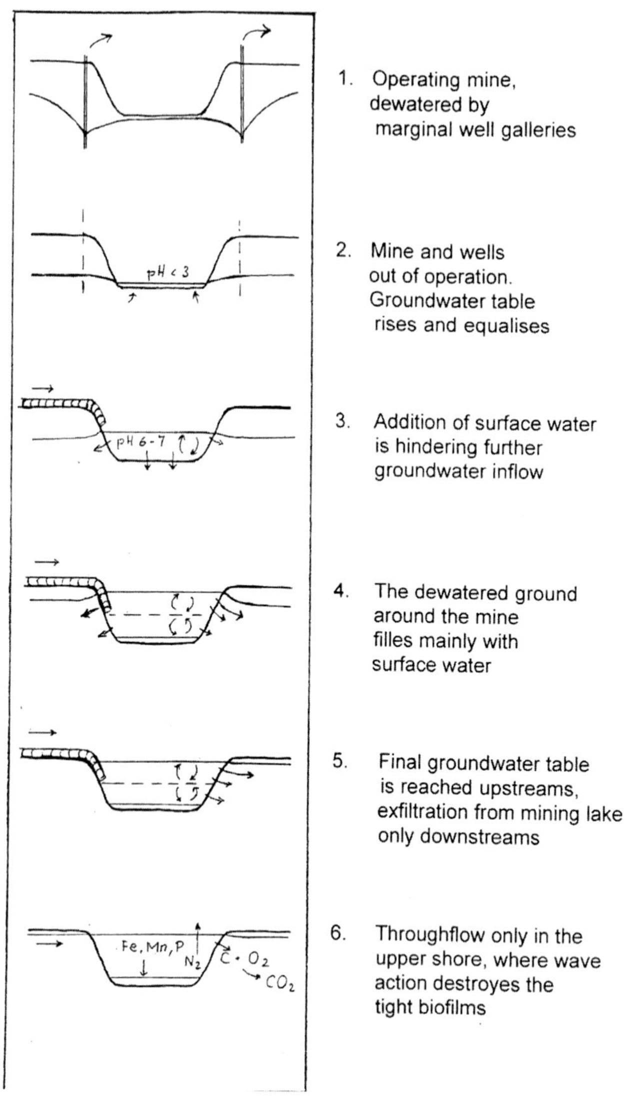

Fig. 10. Filling of a mining hole with surface water; typical sequences (from Klapper and Schultze 1993)

Depending on the nutrient supply of the mining lake in question, the primary production is concentrated in the near-surface layers of nutrient-rich eutrophic lakes or distributed to greater depth in the case of oligotrophic lakes. According to the primary production, the oxygen in the deep water is depleted in eutrophic and nearly saturated in oligotrophic lakes. Very clear waters may be oversaturated at depth, e.g. when *Juncus bulbosus* has overgrown the bottom and performs photosynthesis, as occurs at a depth of 20 m in the mining lake Koschen, Germany.

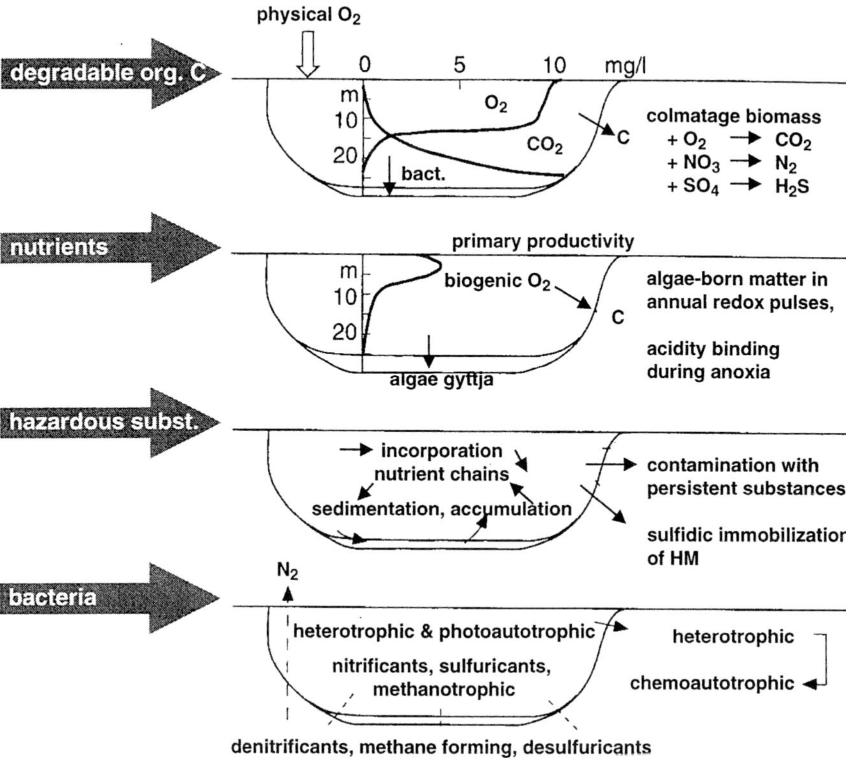

Fig. 11. Different loadings and consequences in mining lakes filled with surface water (from Klapper and Schultze 1993)

Considering contamination with hazardous substances, each case differs from the next. First, the character of the contaminant has to be established: inorganic or organic, degradable or not, solubility, possible introduction in nutrient chains, etc. Groundwater protection deserves high priority, especially against non-degradable contaminants and substances being toxic during bioaccumulation. Nevertheless, the first question should be whether a deposit of hazardous substances exists and what special conditions have to be established for functioning degradation and to allow the restoration of the contamination by in-situ self-purification (see Chapter "Loading of Mining Lakes and Water Quality Control. 3 Contamination"). Bacteria communities in mining lakes differ from those in natural lakes particularly in the first filling stages. Chemolitotrophic iron-bacteria and sulfur-bacteria play the most important role. Bacteria and microalgae hardly compete for the very limited resources of carbon and nutrients. Because of the scarcity of carbon and phosphorus in young mining lakes, such species of algae seem to have selectionary advantages, by which they gain these essential resources by ingestion of bacteria: mixotrophic algae with a special adapted metabolism (Tittel and Zippel, pers. comm.).

The microbial food web in mining lakes (Fig. 11) is the subject of current research in laboratories of several countries. At present it is known that it takes at least decades and sometimes centuries to achieve identical conditions in mining lakes as those prevailing in natural (glacier- or tectonic) lakes. Limnological knowledge, standards and other tools for a good water quality management continue to be developed.

5
Sediments and their Role for the Abatement of Acidification

Sediment profiles and the detailed investigation of the undisturbed layers with the related solid and liquid phases provide an excellent opportunity for reconstructing the history of a lake development. In mining lakes, profiles down to the prelimnic bottom cover the whole succession from the first filling to present day. Therefore, the sediment is correctly called the memory of the lake. With the help of distinct marks, the age of the layers may be determined, so the annual sedimentation rate may be estimated. Such marks include the Cesium peak from the Chernobyl disaster, 1988, or annual white stripes from biogenic calcite precipitation in natural lakes. In some young mining lakes with large amounts of iron precipitated, black and brown stripes allow one to distinguish between the anoxic conditions, due to degradable bioproducts, in summer and oxic conditions in the winter. However, in any one mining lake the sedimentation conditions differ from one area to another. Therefore, the first step is to find correct and representative sampling points. During the first time of groundwater entrance some sediments may be accumulated only in the deepest furrows of the former mine. As long as wind and wave action are able to re-suspend the sedimented matter from the shallow water, the submerged slopes at the lake bottom remain free from sediments, and by the "funnel-effect" the particles concentrate at the deepest localities of the basin (Fig. 12).

Therefore, to estimate the annual sedimenation rate, this variation from no sedimentation at some sites to sediment accumulation at the others has to be considered. The first stage of mining lake formation, i.e. the empty open-cast mine without water, has no sediment at all. The bottom of the hole to be filled consists of sand, gravel, overburden material and residues of non exploited lignite. Important coal areas may be in contact with flooding water, where they were prepared for mining, but not mined due to the new economic conditions.

In Germany, most of the smaller mines were closed immediately after the German unification because of the breakdown of industry and its energy demands. An important question in such cases was the influence of these lignite areas on the water quality of the generating lake and whether it would be necessary to cover the coal with inert soil material in order to avoid occasional water pollution. Fortunately only little effect from the coal on the water column was found and the covering of the coal could be neglected.

The carbon content entering the water as compounds similar to humic substances was much lower than assumed from the red-brown colour of the rain puddles in those areas. Obviously the discoloration originated not from coal extracts, but from dissolved ferric hydroxides at pH < 3. The carbon content in acidic mining lakes has proven to be too small for "normal" functioning of the aquatic ecosystem. With pH < 3 no carbonate, no bicarbonate and only traces of carbon dioxide are present in the sulfuric acid environment. And with low pH-values, low DOC-contents could also be determined. Therefore, it is hypothesized, that carbon is limiting the life starting processes (Klapper and Schultze 1995).

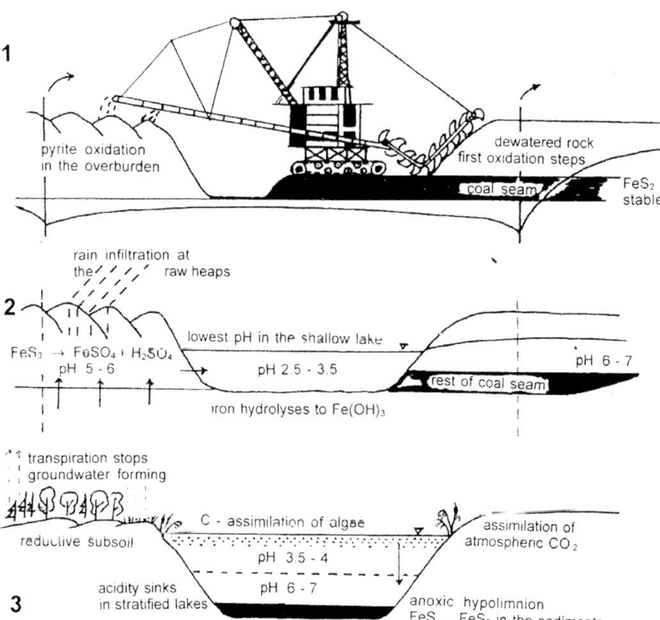

Fig. 12. Development of the lake bottom from acidity source to acidity sink

More important than fulvic substances from lignite, is the organic carbon produced by photosynthesis of plankton algae and the emergent macrophytes. Organic biomass sinking down to the sediments together with iron hydroxide enriches the uppermost sediment layers with degradable matter. Biogeochemical processes in surface sediments such as decomposition of organic matter, reduction of manganese and iron oxides, and bacterial reductions, are the key factors influencing sediment chemistry. Depending on the quality of the organic matter and the lake dynamics, hypolimnetic anoxia can lead to enhanced microbial alkalinity production from anaerobic respiration, raising the pH in the sediments up to neutrality (Wendt-Potthoff and Neu 1998). In particular, iron-reducing bacteria may play an important role in the biogeochemistry of sediments that were influenced by acid mine drainage (Bell and Mills 1987).

The importance of iron-reducing bacteria could be confirmed with help of sediment investigations on mining lake 111 in the Lusatian district (Table 6, Fig. 13, Friese et al. 1998). This lake is 30 to 40 years old, so it can "tell something" about sedimentation processes and diagenesis of the sediments in the past. Mining Lake 111 is used by the Institute of Inland Water Research, Magdeburg, as an experimental lake. The lengths of the sediment cores varied at the deepest parts from 14 to 100 cm. The pre-limnic bottom was easy to be distinguish by the sand and coal particles.

Table 6. Mining Lake 111; main limnological data

Surface area	10.7 x 10^4 m^2
Volume	0.5 x 10^6 m^3
Maximum depth	10.2 m
Mean depth	4.6 m
pH (median)	2.6
Sulfate	1310 mg/l
Iron (median)	156 mg/l

Where the lake possesses an anoxic hypolimnion, the sediments are black from reduced iron. On shallower localities the sediments are more light-brown from oxidized iron. Analysis of the cores from the deepest area allows for some general observations (for details see Friese et al. 1998). The redox-potential measured by punch-in electrodes in the field falls from +600 mV on the sediment surface to zero at 2 cm and to –100 mV at 4 cm and lower until 20 cm. The pH measured in the same core was 2.8 at the sediment-water interface, increased until 1 cm to pH 4, prevailed at pH 4 to 2 cm, increased evenly until 8 cm sediment depth to greater than pH 6 and remained at this value until the greatest depth of the core at 20 cm. The surface represents the iron-buffer, 1 – 2 cm the aluminium-buffer and from 8 cm depth the circumneutral bicarbonate-buffer is governing. With the exception of the liquid surface and the sandy bottom the core was sectioned in 1-cm subsamples and analyzed in detail.

Fig. 13. Research object Mining Lake 111

It could be demonstrated, that during the lifetime of the lake the bioprocesses increased constantly. At the deepest, i.e., from the oldest core section (14 cm), the lowest values were measured in loss of ignition (LOI) 5 %, C_{org} 20 mg/g, TC 20 mg/g, TN 0.5 mg/g, TS 1 mg/g, P 0.2 mg/g. The related values at 2 cm were LOI 22 %, C_{org} 80 mg/g, TC 110 mg/g and TN 7 mg/g dry weight.

The maximum total sulfur was found with TS = 17 mg/g in the 2 – 4 cm section. Phosphorus had two maxima in the 2- and the 6-cm layer with 0.55 mg P per gram dry weight each. The authors assume, that most of the sulfur at 2 – 4 cm depth represents inorganic sulfides, generated during dissimilatory sulfate reduction by anaerobic bacteria. At this depth degradable organic carbon should be present, the E_h was below –100 mV and oxygen penetration can be excluded. Additional support for this hypothesis comes from the high Fe(II)-content and from the fine-grained black material suggesting the presence of ferrous sulfide.

One target of the sediment research is to find techniques for enhancement of the processes of biological binding of acidity, sulfur, iron and other heavy metals in form of sulfides. For oxygen-free conditions, degradable substrates are needed, whether allochthonous introduced or autochtonous produced. However, economical and ecological difficulties are connected with both.

6
End Use/Costs

From an esthetical viewpoint a reclaimed and well functioning post-mining landscape may be more beautiful and pleasant than that before mining. The first heaps from opening the opencast mine dumped on the ground later form forested hills. Deep holes remaining from the excavated coal are later filled with water. They become new lakes in countrysides, where lakes were extremely rare or unknown before. In some regions the mining lakes are occupying large areas of the landscape.

These water bodies are the most characteristic parts of the newly forming lake districts. Their existence should not be understood as a land loss, but as an opportunity for the future. With appropriate management, these lakes will be usable for bathing, diving, fishing etc. Bank filtrated waters are suitable for drinking and processing purposes. The need to make use of the waters depends on the actual demand and the existing infrastructure. The brown-coal industry has had to supply the whole area with long-distance pipe systems for dewatering of the landscape before the mining started. In eastern Germany, the closure of the lignite mining industry with the mines themselves, the coal power plants, briquette factories, smoulderies and chemical coal processing factories has left a landscape of oversupplied infrastructure with respect to traffic, energy, water, food, etc. So, on the one side, the landscape after mining has an excellent infrastructure, which calls for a secondary utilization by new enterprises and business, but, at least temporarily, there is also high unemployment, with only little chance for a quick change. The whole region has been monostructurally oriented on coal and coal processing for decades.

Considering this special situation, it is not surprising that so much interest is directed on the lakes as possible nuclei and attractions for water-oriented tourism including fisheries. Interviews with the inhabitants of the villages around the new emerging lakes have produced similar results everywhere. The people of even the smallest villages want in future the lakes to become "centers" for recreation in order to promote jobs and salaries in related services. However, for economic reasons, only a few such centers should be planned and established. For decision making, the factors of suitability of each lake have to be evaluated.

The development of a mining lake for recreational purposes at first demands several million German marks or U.S. dollars. The return from this starting investment may be realized only from a relatively high frequency of tourists, spending their holidays at the lake, using the local services and spending enough money in the region.

From an economic point of view the lakes in areas under nature protection are far less interesting. With the orientation on "silent recreation" they will attract nature friends and walking tourists with backpacks and binoculars, carrying their own food and usually representing only a very modest business to the local restaurants.

Decision criteria for economic planning can be listed in three groups:

A natural fit of the lake landscape;
B development for recreational purposes; and
C restrictions due to environmental and other impacts.

To assess the suitability of a lake landscapes for recreational purposes, mean conditions may be evaluated as "1", excellent recreational suitability with "2" and absolute insufficiency, e.g. a closed lake, with "0" (Table 7).

Table 7. Factors for evaluating the suitability of lake landscapes for recreational purposes (adapted from Klapper 1995)

A	**"Natural" fit of the landscape and the mining lakes**	
Aa	Size and variability of the lake landscape	1.0 ... 2.0
Ab	Relief forming of the lake landscape	1.0 ... 2.0
Ac	Recreational value of the broader surrounding	0.5 ... 2.0
Ad	Forming and esthetical value of the shore landscape	0.5 ... 2.0
Ae	Geogenic acidity and iron contents	0.1 ... 2.0
Af	Trophic state (polytrophic ...oligotrophic)	0.2 ... 2.0
Ag	Bioclimatic conditions	0.5 ... 1.5
B	**Development of the landscape for recreational purposes**	
Ba	Demand for leisure-, weekend- and holiday-recreation	0.5 ... 2.0
Bb	Development of the local and long-distance traffic, equipment with parking places	0.5 ... 1.5
Bc	Development and mine-technical safety of the shores	0.1 ... 2.0
Bd	Bathing possibilities (not observed or guarded strands and lawns)	0.2 ... 2.0
Be	Restaurants, camping and sporting sites, water and food supply, waste disposal	0.6 ... 2.0
Bf	Possibilities for further activities of water-bound recreation (surfing, diving, sailing and adequate training schools), water travelling, ferries, angling	0.2 ... 2.0
Bg	Cultural and educational offers (museums, concerts, cinemas, discotheques, nature trails)	0.8 ... 1.6
C	**Factors restricting the recreation**	
Ca	Unhygienic conditions, bacteriologically insufficient surface waters for filling	0.2 ... 1.0
Cb	Utilizations adverse to recreation (ash flushing, dumping of hazardous substances)	0.1 ... 1.0
Cc	Turbidity from iron hydroxide, visible scums of oil, tar, hydrophobic coal dust, floating wastes, wood, plastics, bottles etc.	0.1 ... 1.0
Cd	Closing of landslide- endangered shores	0.1 ... 1.0
Ce	Loading by noise and dust from neighboured open pits and power plants	0.2 ... 1.0
Cf	Overloading by too much visitors	0.1 ... 1.0
Cg	Proximity to wasted factories, production ruins, visible traces of landscape destruction	0.1 ... 1.0

The computation of the recreational suitability, R_S, of a distinct landscape with mining lakes can then be performed as follows. The arithmetic means of factors A and B are added and the sum divided by 2 and multiplied with the mean of the factor C. Evaluation of the recreation suitability R_S allows decisions as to which alternative recreational projects are worth subsidising (Table 8). The value of a restoration scheme may also be expressed as an increase in the recreational suitability.

Table 8. Evaluation of recreational suitability

R_S			assessment
0	...	0.1	not suitable
0.1	...	0.5	partly suitable
0.5	...	1.0	suitable
1.0	...	1.5	well suitable
1.5	...	1.8	excellently suitable

A capacitive figure, the recreational efficiency R_E, may be introduced. It is given by multiplying R_S by the recreation days R_D, being realized at the mining lake:

$$R_E = R_S * R_D$$

One recreation day may be defined as at least two hours stay of one person with realization of at least one water-oriented activity. To achieve a monetary expression, the R_E has to be multiplied with the well-known costs of one recreation day in an artificially built open-air bathing basin (share of building costs, water treatment and water exchange from public supply, services etc.). This cost may be different in diverse countries, but in every case it is usually much higher than the cost in the new emerging mining lakes, which are large enough that bathing water treatment is not necessary. With this substitutional method only the possibility for bathing is evaluated. The comparably higher value of the new mining lakes results from the potential for diverse water-oriented activities, from the size, the self-purifying large water body, the connection with other lakes and further recreational possibilities in systematically developed landscapes for leisure purposes.

The general question remains, as to whether the mining industry has fulfilled its duty once the lakes are filled and the banks are safe against sliding, or must the lake be usable and the quality of the water suitable, before the lakes may be handed over to the public.

For most cases in Germany, the usability is included into the duties of the mining companies. In the new federal provinces, special enterprises have been established for realizing this. In a countrywide context, this task is not only important for the unemployed people in the mining region itself, but also for the densely populated greater regions of Berlin, Cottbus, Dresden, Leipzig and Halle. The traffic connections generally allow access to the new lake-lands in less than one hour by car or somewhat more by public transportation. The necessary costs have to be invested in the interest of the society – the former mining employees, now waiting for jobs in service enterprises of the tourism industry and also the people from the large cities, wanting to regenerate their working power by active, water-oriented recreation.

One reason for improving the water quality is the fishery. The special conditions in the larger clear water lakes are appropriate for an occupational fishery on the basis of the pelagic food resources. Provided a pH constantly greater than 5.5, or better 6.0, *Coregonus albula* may utilize the zooplakton. Because of the low level of gross bioproductivity in mining lakes, it is useful to produce precious fish species. This way a good monetary yield might be gained with only a small weight of harvest. In larger mining lakes further fish species like pike (*Esox lucius*) and eel (*Anguilla anguilla)* should be inserted, in addition to the introduced and natural fish stock. This constitutes a form of biomanipulation, i.e. the top-down-control of the food chain. Instead of the ichthyoeutrophication to be observed with some cyprinid species a positive influence of the fishery on water quality is possible this way. The best monetary income at such mining lakes with the priority of recreation follows from the sale of licences for sport fishing (Klapper 1998a).

Summarizing this chapter, it can be realized that a usable mining lake has a good potential for a return of the costs that are required to provide safe banks and a sufficient water quality. Moreover, with a diverse selection of mining lakes in the same region, the landscape concerned should develop within a relatively short time as an attraction within the tourist industry. Many people may get new jobs in this service sector of the regional economy.

Lakes established with the only target of being interesting features in nature-near landscapes, having a natural succession without human assistance, may be relatively cheap. However, related to this low expense there will be nearly no financial return or economic benefit to the population. Of course, in the sense of nature protection and for limnological science the value is very high, but it is not acceptable to express natural heritage with its species diversity and natural services in a monetary form.

7 Acid Mining Lakes

7.1 Do you Need to Treat?

Geogenically acidified mining lakes are relatively unique objects of limnology. A few volcanic lakes in Camchatka and Japan and some smaller examples like the sulfur acidic Tonteich Reinbeck near Hamburg, Germany, are described in the limnological literature (Ohle 1981). The largest of such extremely acidic lakes are presently emerging in the Lusatian mining district of Germany.

Without intervention, the lakes would stay acidic for decades or centuries unusable for water-oriented recreation with primary body contact. Also, water sports with secondary body contact, such as canoeing, sailing, etc., would be restricted. Fishing related activities, such as angling and professional fishery, would be absent from these fish-free lakes.

Because many smaller lakes from the older, smaller mines exist, there is no lack of examples for studying all of the problems connected with geogenical sulfur acidification. Such lakes are actually also valuable as part of nature-protected landscapes. In the interests of nature protection some overburden heaps have been kept without recultivation, to become the so-called succession areas, where the vegetation remains very scarce, without trees and with no closed grass cover. Such areas have become places with relatively hotter microclimates within otherwise moderate climatic zones. Hence, these areas are rich in plants adapted to dry conditions, coloured flowers, rare species of insects, and reptiles, etc., usually more common in warmer areas. The groundwater under these heaps remains oxic and acidic. The unrecultivated territory is thus connected with long-term acidification of the lakes in these areas. Eventually, in the future, the acidification will come to a natural end in these areas. This will occur once all oxidized pyrite has been washed out or when the trickling rain water has washed so much of the sulfuric acid to greater depth that a closed vegetation cover will occur and bioproducts and roots will degrade. Further, oxygen in the groundwater will be depleted and the washout of sulfuric acid will stop due to anoxic conditions and reduction of sulfur and iron to form iron sulfides.

Regarding the large lakes emerging from the most recent opencast mines, there is no doubt that only restored, neutral lakes have the quality to be released from the responsibility of the mining enterprise for utilization by the public. In the past, mining engineers had a tendency to consider the mining lakes ready for release after achieving sufficient stability of the banks. However, there was no recreational interest in such useless, large lakes filled with dilute sulfuric acid. The mine enterprises would typically be stuck with ownership of these non-marketable lakes for a long time, and ultimately, this would be more expensive than to satisfying the financial requirements for neutralisation.

From a socio-political viewpoint the necessity to reclaim past surface mining areas is obvious. The people in the mining districts have the right to make decisions about the future of the region, in which they have been working and where mining companies received much money by exploiting the resources and temporarily destroying the landscape. The excellent infrastructure in past mining areas should attract new industries and a beautiful landscape, with lakes, should attract holiday makers. Many jobs in the tourism-sector may be found, provided that the lakes in the region would be usable. With these points in mind, the following chapter gives a brief overview of the necessary steps for abatement and control of acidificaton of acid mine lakes.

7.2 Methods of Curbing Acidity During Lake Formation

Alleviation of acidification should be considered and implemented right from the initial planning stages of mining operations, up until the recultivation of the overburden heaps. If sustainable neutralization is to be obtained, a complex, long-term abatement program must be set up covering a broad range of issues.

In order to minimize the contact of pyritic minerals with oxygen, the temporal and spatial dewatering of the lignite to be mined must be limited. The layers of overburden with the highest pyrite levels should also be dumped in the deepest part of the mine, thus enabling these materials to become submerged as early as possible. The diffusion coefficient for oxygen in water is only 0.0001 of that in air.

The final position of an abandoned mine in the groundwater field influences the acidity input into the mine lake. Groundwater from the undisturbed rock is mostly neutral, whereas that from the overburden heaps is acidic as a result of the oxidation of all the sulfidic ores during the mining process. Therefore, the end-position of an abandoned mine in the greater groundwater field should preferably be downstream of the undisturbed rocks and upstream of the groundwater forming in the overburden heaps.

The important role of the filling water has been already described in Section 2. Mining lakes, extending over great distances in the direction of groundwater flow may act like discharging or recharging groundwater wells. The lakes are horizontal levels within the slope of the underground water table. A deeper lake level draws water from surrounding aquifers upstream, while lower aquifer levels cause surface water to infiltrate down through the ground beneath the lake. In order to avoid intensive flow and mass transport of dissolved matter, such long-shaped lakes should be subdivided, thereby decreasing the hydraulic gradient, as shown in Fig. 14.

Revegetation of the acidic heaps can be boosted by chemical treatment with alkalinity-delivering materials such as lime or ash. In eastern Germany, with low precipitation of about 500 mm/year, the establishment of mixed forests, with the associated high evapotranspiration rates, reduces groundwater infiltration to nearly zero. The soil and humus cover of the forest decreases unwanted oxygen penetration to the subsoil (#1 in Fig. 14). Similar findings are to be observed with permanent grassland and organic fertilizers (#2 in Fig. 14). Wetlands and fishponds can be reestablished in the lowlands along the rivers and creeks. These were typical of the Lusatian landscape in Germany prior lignite mining. The wet covers prevented the aeration of the ground (#3 in Fig. 14). The application of detergents, which kill the bacteria responsible for acid formation, is appropriate for smaller heaps rich in sulfidic ores. Although toxic to *Thiobacillus,* these materials provide a degradable and oxygen-consuming substrate for the other heterotrophic microorganisms (Kleinmann et al. 1981; Rastogi 1996; #4 in Fig. 14).

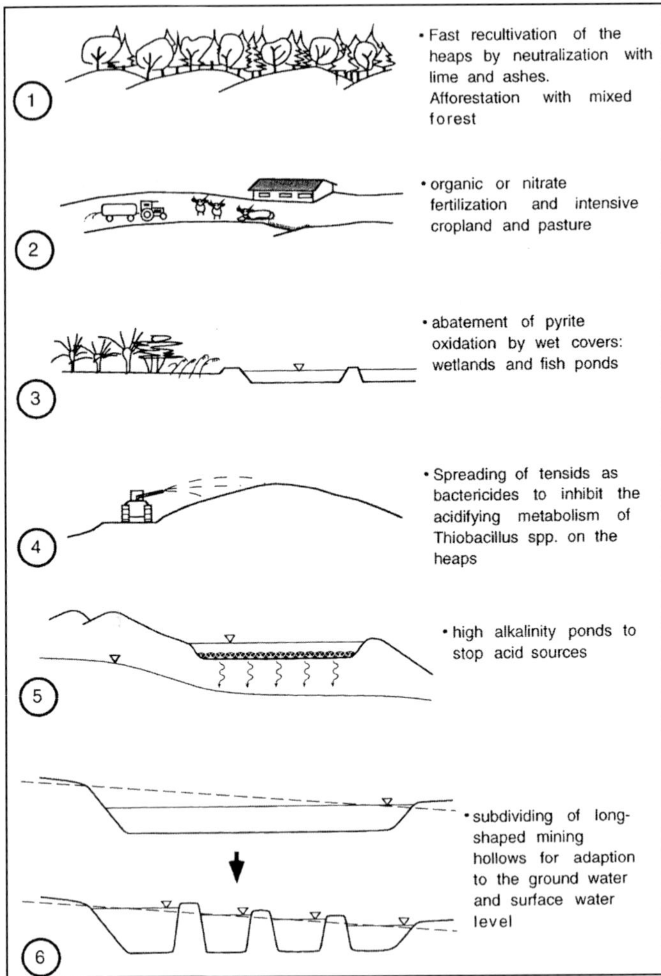

Fig. 14. Abatement of acidification by measures in the drainage basin (from Klapper and Schultze 1997)

High-alkalinity ponds are recommended for neutralizing ground-water outlets to mining lakes. With limestone on the bottom, such ponds promote infiltration of alkalilnity into the acidic ground (#5 in Fig. 14).

7.3 In-situ Neutralization of Mining Lakes

The natural geogenical processes causing acidfication are summarized in Fig. 15. The left side corresponds to conditions in a lake, filled only with groundwater and being too shallow to become stratified by temperature-driven density gradients. With oxygen present in nearly all compartments of the lake system, including bottom sediments, the acidity will stay for decades. On the other hand, in a stratified lake (right side) there are anoxic parts in the sediments, in the hypolimnion and, in cases when the lake is meromictic, in the monimolimnion. Carbon-limitation may be overcome with external carbon imports or by bioproduction. The emergent macrophytes are independent with respect to carbon-supply, because they receive CO_2 from the inexhaustible resource in the atmosphere. At the same time they produce degradable organic material by photosynthesis. The sediments of the reed belt, with its rhizomes, are areas of sulfate reduction, with the visible black colour of the rotten plants, sometimes interspersed with golden crystals of pyrite. In addition, leaf-litter from trees near the shore produces microhabitats for sulfate-reducing bacteria.

Fig. 16 illustrates a number of ecotechnological in-lake approaches for neutralization. The ideas originate from observation, the literature, experience gained while tackling similar problems (e.g., with the aim of controlling nitrate), and from laboratory experiments. Many relatively small and shallow lakes have remained acidic for decades. Other lakes with stable stratification were neutralized within few years by internal mechanisms. Ecotechnologies functioning on the basis of microbial sulfate reduction require the absence of oxygen and nitrate, and the availability of a carbon source for sulfate respiration (Fig. 15). Both stratification and mixing in a lake depend on the length of free wind action or the wave fetch. Stratification is a desired condition for acid mitigation.

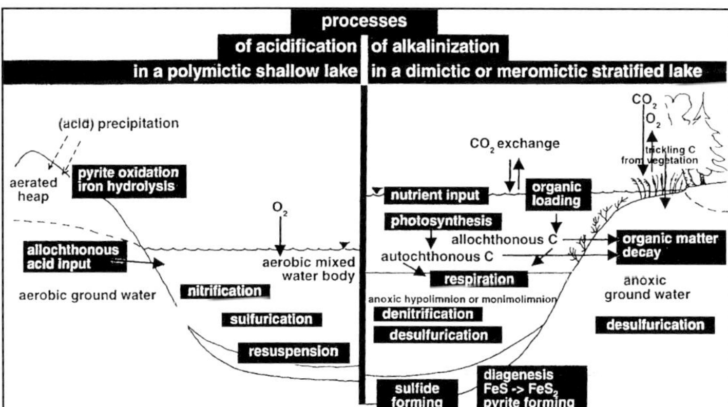

Fig. 15. Processes affecting acidity in shallow and in stratified water bodies (from Klapper et al. 1998)

To show how efficient stratified systems can be, Lake Fuchskuhle, a shallow moor lake in eastern Germany, was divided into four limnocorrals with help of plastic sheeting. The corrals became stratified with anoxic deep water and sediments. As a result, the pH changed to nearly neutral while nutrients and bioproductivity increased (Babenzien 1996).

Possible tools to induce such stratification in mining lakes include oil barriers, floating reeds and submerged foils positioned from the bottom into the open water like a fishing net. The provision of wind protection by the shore is only effective where lake areas are small (#1 in Fig. 16).

The main way of attaining a neutral mining lake in the Lusatian region of Germany is to first rapidly flood the pits with surface water. Although installation of the necessary pipelines or ducts is expensive, this approach has the additional benefit of providing shore stability (Luckner et al. 1995). Running surface waters contain neutralizing and buffering bicarbonate, nutrients, and dissolved and particulate organic matter, all of which are ingredients for the abatement of acidification. In order to avoid excessive nutrient input, towards the end of the flooding process the river should function only as a bypass of the lake. To satisfy any neutralization requirements, no more than the necessary amount of surface water should be used (#2 in Fig. 16).

The addition of phosphorus to a rain-acidified, soft water, upland lake in England led to neutralization by the above-described mechanisms. A total of 5.9 m^3 of a phosphate solution brought about the effect of 34 tonnes of calcium carbonate (Davison et al. 1995, George and Davison 1998).

For restoration of the pelagic food web in acidified and limed lakes, a gentle fertilization may be useful (Bell et al. 1993; Olofsson et al. 1988; #3 in Fig. 16). In very acidic mining lakes, fertilization with treated sewage containing not only nutrients but also organic carbon appears to be successful. One good example of neutralization by sewage is the case of Laubusch mining lake in Germany. During stagnation periods sulfur and iron are transported into the black bottom sediments, fixing the acidity. With pH values of 3 – 4 in the epilimnion and 6 – 7 in the hypolimnion, a good stock of fish has developed in this lake (Klapper and Schultze 1995). In lakes, neutralized by flushing with surface waters, but endangered under low water conditions by entrance of acid groundwater and reacidification, a desirable eutrophication may be achieved with the help of trout breeding in net containers. This will help to stabilize the neutral conditions with generating anoxic sediments and deep waters. Desired oxygen consumption results from uneaten food and fish feces. This kind of artificial eutrophication is not connected with pipelines or other construction and may be licensed temporarily, so that it may be finished, if necessary, relatively soon. For the professional fisheries this kind of fish production is profitable and would stabilize the income in those cases, where the *Coregonus* (white fish) harvest is too low for a sufficient income. This has been realized at the mining lake Senftenberger See in Germany (Rümmler, pers. comm 1998).

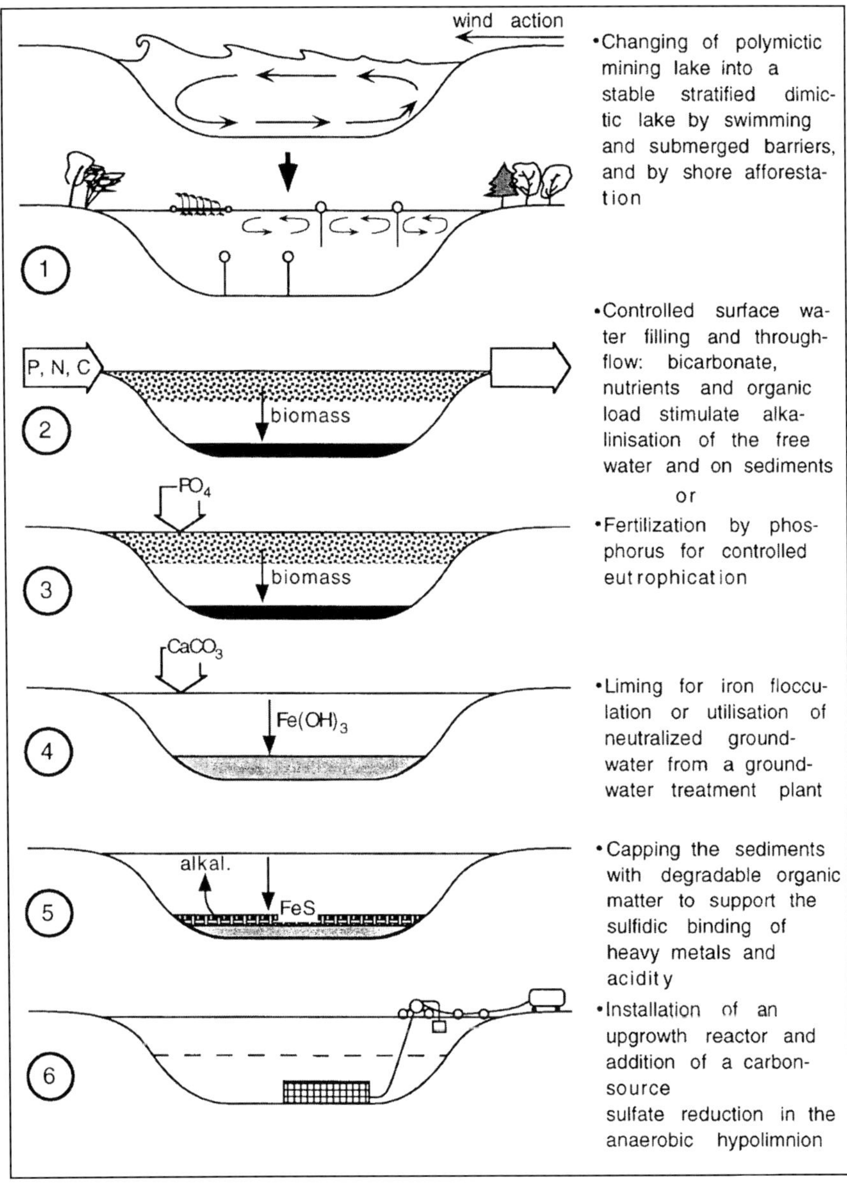

Fig. 16. Abatement of acidification by in-situ technologies (from Klapper and Schultze 1997)

Liming has been successfully applied in Canada, the USA and Sweden to neutralize rain-acidified lakes. In Sweden, about 5000 surface waters have been neutralized since 1988 under a government program (Olem 1991). In most cases calcium carbonate is used, providing alkalinity and inorganic carbon. The soft-water bodies are quite different from hard-water mining lakes. The latter are strongly buffered by iron at a far lower pH and the alkalinity demand is about tenfold higher for neutralization. In the Lusatian region of Germany several groundwater treatment plants exist for the neutralization of effluents. They may now be used to produce limed and cleaned water to fill newly emerging lakes. However, this ex-situ treatment is expensive (#4 in Fig. 16), consequently restoration research focuses on natural process of alkalinization by desulfurication.

The addition of organic matter may provide a way of enhancing the sulfide-forming properties of the sediments. Promising experiments are currently being performed by Fyson and Nixdorf (1996, pers. comm.) with potatoe chips as the carbon source, as well as by Frömmichen (1998, pers. comm.) with the addition of organic loaded saturation chalk from sugar factories or liquid organic substrates. Another proposal has been to dredge the sludge from eutrophic lakes and to spread such lake sediments on the bottom of acidic mining lakes. This method of restoration is designed to counter eutrophication in the case of natural lakes and acidity in mining lakes (Nixdorf et al. 1997; #5 in Fig. 16).

Another example of an effective anaerobic ecotechnology has been developed for heterotrophic nitrate dissimilation in Zeulenroda Reservoir, Thuringia, Germany. A steel cage measuring 20-m x 60 m x 1.5-m, filled with 13000 bales of straw, and equipped with distribution pipes, was positioned on the hypolimnic bottom near the dam. Nitrate-rich surface water was pumped through the straw reactor together with fatty acids to provide a carbon source. The dissolved oxygen was quantitatively released, after which the nitrate oxygen was utilized. Nitrogen escaped in molecular form. If methanol is used to represent the hydrogen donor, the process can be expressed by the following overall equation:

$$6\,NO_3 + 6\,CH_3OH \Rightarrow 3\,N_2\uparrow + 6\,CO_2 + 12\,H_2O^-$$

This technique enabled the hypolimnion to become anaerobic and nitrate-free within eight weeks. The water was reaerated along a stretch of 5 km of turbulent flow between Zeulenroda and a terminal reservoir and proved to be a suitable source of raw water for drinking water supply (Fichtner 1983; Klapper 1991; #6 in Fig. 16).

7.4 Ex-situ Treatment of Mining Lakes

A pilot scale, ex-situ bacteriological sulfate reduction scheme is underway at the mining lake Kahnsdorf south of Leipzig, Germany (Glombitza and Madai 1996). The closed anaerobic biofilm-reactor is supplied with lake water and methanol is used as a carbon source. Methanol is needed at a ratio or around

1:1 with reduced sulfate. Vibrionic bacteria, immobilized on up-growth carrier material, are most efficient. Methanol is consumed totally. A drawback is the relatively low capacity for the large volume of the lake (see also Brettschneider and Pöpel 1992).

A further approach is based on an electrolytic hydrogen generation, with reduction of oxygen and precipitation of heavy metals. By this approach the pH rises by proton consumption. The energy demand is preliminary estimated to be 0.5 – 1.2 kWh/m^3 lake water (Friedrich 1996). Application of this technology on a technical scale has yet to be established.

Several established ex-situ technologies for the recovery of acidic waters are summarized in Fig. 17. Alkalinity production or acidity binding is best accomplished in fully or partly anaerobic systems, while the precipitation of unwanted heavy metals is best achieved in aerobic macrophyte systems after the pH has risen to neutrality. Many proposals comprise combinations of anaerobic with aerobic steps, sometimes with limestone as an inherent constituent of the various stages. For example, anoxic limestone drains are widely used in the U.S. to satisfy the alkalinity requirement of acidic and metal-containing mine effluents. Anoxic operation is necessary to avoid clogging by the precipitation of metal hydroxides. A trench is dug and lined with plastic sheeting before being filled with limestone. The ends of the plastic sheeting are wrapped over the limestone and covered with soil. The inlet and outlet are protected against aeration by traps such as u-bend pipes (Hedin and Watzlaf 1994, #1 in Fig. 17). This approach may also be suitable for the lignite-mining regions of Germany. As soon as a self-sustaining water balance has been re-established, the neutralizing service ecosystems (i.e., anoxic limestone drains) should be set up at lake inlets or along connecting trenches between two mining lakes.

Several other technologies have potential application to the East German lignite-mining regions. For example, in California, deep, narrow trenches filled with bales of straw serve as anoxic denitrification facilities (Brown 1971; Sword 1971; Jones 1974). When denitrification is complete desulfurization takes place in the same system (#3 in Fig. 17).

At the Wheal Jane Pilot Plant in the UK, closed plastic-coated upgrowth reactors are filled with degradable or inert up-growth materials and, with the addition of substrate, support microbial desulfurization. Sulfides remain inside the reactor, but the effluent has to be re-aerated (Taberham pers. comm.; Lamb et al. 1998; see #4 in Fig. 17).

Successive alkalinity-producing systems can also be used. These consist of infiltration ponds with a drained layer of limestone covered by an organic layer, put in place to consume the dissolved oxygen. However, problems with throughput can arise from clogging of the bottom by microbial biofilms (Kepler and McCleary 1994; Nawroth et al. 1994; #4 in Fig. 17).

Anoxic biofilm chambers, which are deep small ponds supplied with oxygen-consuming organic matter, operate without infiltration.

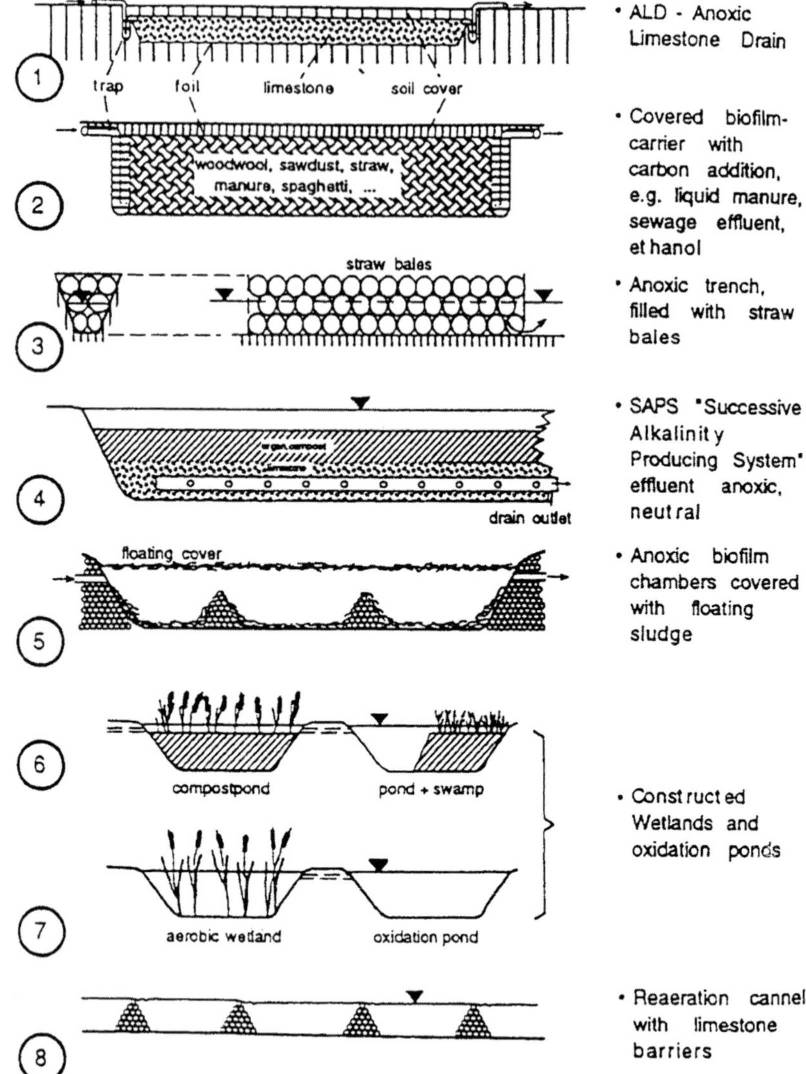

Fig. 17. "Service ecosystems" for neutralisation of acidic effluents (from Klapper and Schultze 1997)

Comparable to inefficient sludge digesters, the floating cover in this case is an essential part of the technology. Silage fodder has proven to be the most suitable organic material (Kalin pers. comm.; Philipps and Bender 1998). The mining lake Haselbach 1, south of Leipzig, Germany, has by chance developed in this way. The 6-ha large, 10-m deep lake is completely overgrown by Cattails *(Typha angustifolia, T. latifolia).* Its waters have been used in a briquette factory and as a settling basin for cleaning waters of an industrial fodder drying facility. There were sufficient hydrophobic floating coal particles and heterotrophic biofilms for the cattails seed to sprout and to overgrow the whole mining lake within two years. Now the lake is a nature protected area and it is strongly forbidden to walk on the dangerous floating plant cover. At present, artificially grown reeds on special mats made from coconut fiber are available for starting such floating macrophyte covers. Under the floating cover the water is black, anoxic and neutral, but hidden from sight. At first glance the reed looks like a pleasant wetland (#5 in Fig. 17).

Constructed wetlands, in various configurations, and oxidation ponds serve the aerobic polishing of anaerobic neutralized waters (#7 in Fig. 17). A few approaches include anoxic sections (#6 in Fig. 17). Metal hydroxides are transferred into the permanent bottom sediments (Hedin 1989). For successive alkalinity delivery, watercourses may be equipped with limestone beds. Reaeration can be implemented with the help of limestone overflow barriers (Dietz et al. 1994; #8 in Fig. 17).

Summarizing this section on acidification, it appears that the abatement of geogenically sulfur acidification includes measures to combat pyrite oxidation, steps to decrease groundwater and acidity transport, as well as in-situ and ex-situ neutralization processes. Chemical neutralization is often impracticable because of the large amounts of alkalies and the costs of the treatment. Large mining basins should primarily be flooded with surface water containing bicarbonate. A temporarily higher trophic level in the lakes must be tolerated. The most promising alternative for chemical neutralisation is the encouragement of microbial processes of anaerobic acid binding by desulfurization. Aerobic treatment with macrophyte systems, such as in artificially constructed wetlands, is suitable for completing water treatment via the flocculation of the heavy metals contained as hydroxides. The research findings discussed above can in principle be applied to the mining of nearly all pyrite-containing rock such as hard coal, quarz and slate, as well as sulfidic ores (especially uranium). More detailed information about the basics of geogenic sulfur acidification, limnology and management is given in the book "Acidic Mining Lakes" by Geller et al. (1998).

8 Eutrophication

8.1 When to Treat?

Control of eutrophication is high priority in neutral mining lakes, in order to achieve a good water quality. However, definition of "good water quality" depends on the final use of the lake. As outlined before, water quality is usually assessed from the standpoint of utilization by humans. High demanding uses, e.g., for bathing or drinking purposes, need clear water with a low bio-productivity, i.e., low content of plankton algae. Nevertheless, in the case of shallow mining lakes in protected natural landscapes, the aim of which is high species diversity, eutrophication is no drawback and control of eutrophication should not be a necessary target of management. Consideration of acidic lakes where a distinct eutrophication may contribute to the formation of anoxic parts in the lake for generating microbial binding of acidity, has been described in the above chapter.

8.2 Prophylactic and Dietary Measures Against Eutrophication During Lake Formation

Water quality management of neutral mining lakes should be performed with the help of the knowledge and instruments that have been developed by the limnological research community for neutral lakes and reservoirs. However, in connection with mining lakes there are often many further opportunities to utilize the benefits of the generated water quality. In Section 1, the hydrographic principles required for the formation of clear water lakes were summarized: great depth, steep banks, large hypolimnia, small drainage basin etc. Unfortunately, the depth is often decreased by dumping of overburden layers from other mines or flushing the ash from power plants into the empty mine, instead of creating heaps or ash ponds above the surface of the old land.

Mining geologists responsible for the safety of the banks and shores of mining lakes, work at great expense to generate shallow slopes with a low probability of sliding. From a limnological viewpoint, flattening of the shore decreases the volume of deep water, increases the area of lake bottom in the epilimnion, decreases the mean and maximum depth, etc. Nearly all morphometrical factors influencing the trophic state are worsened. This conflict between geo- and hydro-scientists has to be solved as a compromise. Slopes have to be reduced where the safety is endangered, but not more than necessary. There are also possibilities to hinder people from entering the bank-side by planting thorny brambles, or similar shrubs and trees. Some steep slopes should remain for ecological reasons. At stretches undisturbed by mining, where rock layers have good stability, vertical shores are desirable as they provide valuable habitats for rare animals.

As discussed above, water quality is already influenced during the formation of an opencast mine. When the water has completely filled the hole, the highest priority has to be given to the control of the main cause of eutrophication, i.e., to the nutrient input.

Phosphorus is the most important growth-limiting factor in neutral mining lakes. Because of the special conditions by which phosphorus is precipitated by iron-rich groundwater inflows, and due to P-binding on the mineral bottom of recent mining lakes (and given a relatively high nitrogen supply), *a decrease of phosphorus concentration in the lake water is the key for controlling eutrophication and improving the water quality*. The spectrum of possible measures to control eutrophication will be discussed with help of a simple illustration, developed for neutral lakes (Fig. 18).

Items #1 and #2 in Fig. 18 are interesting in the case of mining lakes. A sewage-free drainage basin may be best realized by sewage diversion to more distant sewage treatment plants outside of the drainage basin. This concept of greater treatment plants includes also tertiary purification by chemical phosphorus precipitation or by biological P- and N-elimination by the anaerobic/aerobic change within the activated sludge process. For example, in Germany, nutrient elimination is prescribed for all larger treatment plants serving more than 20,000 inhabitants or equivalent.

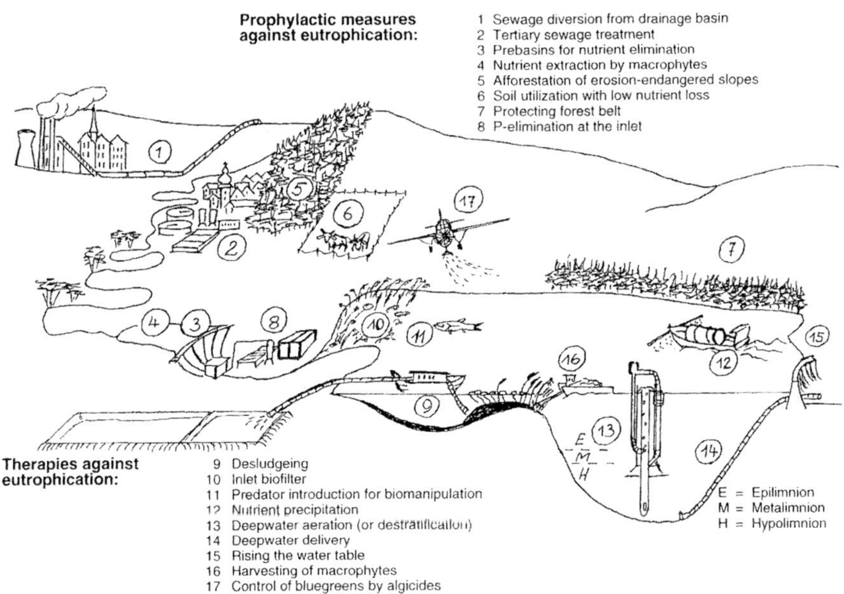

Fig. 18. Methods to protect and restore lakes against eutrophication (adapted from Klapper 1980)

The utilization of storage reservoirs for nutrient elimination (#3 in Fig. 18) has been discussed in Section 2, highlighted by the case of the Mulde Reservoir. Techniques for harvesting nutrients by macrophytes are relatively modest and are restricted to the growing period (#4 in Fig. 18). Protected, shallow underwater parts in the inlet area have to be prepared for this technique. When the final water level is reached and the shore and bottom are sufficiently stable, it is possible to develop *Typha*- and *Phragmites*- stands by systematic planting. The macrophytes are the up-growth carrier for the periphyton, consisting of biofilms with autotrophic and heterotrophic microorganisms. Macrophytes and periphyton form biofilters or the so-called bioplateaus (Oksijuk and Stolberg 1986).

Reforestation is part of the overall recultivation of the mining territory and in particular the heaps (#5 in Fig. 18). In the context of the mining lakes, the importance of reforestation is the associated high rate of evapotranspiration, which curbs the rate of groundwater infiltration. Forestation of reclaimed mining areas is discussed in Chapter "Alternatives for the Reclamation of Surface Mined Lands. 3.3 Forestry".

Reducing nutrient export from agricultural land uses is one means of tackling the dangers of eutrophication in mining lakes (#6 in Fig. 18). To achieve this, well-known principles from soil science have to be followed to establish stable soil ecosystems. A good, crumbly structure and sorption ability will emerge, when soil is well supplied with organics, settled by bacteria, fungi, collemboles, worms, etc., and carries a vegetation cover the whole year.

Protecting wood strips are shown in Fig. 18 (#7) as coniferous forests. These have been recommended for drinking water reservoirs, to avoid fallen leaves loading the water body. However, use of coniferous trees alone should be restricted to the edge of the wood along the shore. It is recommended that the rest of the forest should be cultivated with mixed deciduous and coniferous forests. With leaf-mould, such forest produces a healthy soil fermentation with good sorption ability. On the contrary, pure needle deposition produces a non-degrading raw humus with a poor sorption capacity for key nutrients.

Phosphorus elimination at a lake inlet (#8 in Fig. 18) has been used in research and planning stages, to treat unsuitable mine-filling water from loaded rivers. Since such treatment plants are only used at one location for a few years, they should be designed with:

- robust and simple technology,
- a small percentage of non-transportable concrete buildings,
- a high percentage of prefabricated units, of variable applications based on a modular principle,
- low maintenance and operation expenses,
- simple deposition of flocculation or adsorption sludge,
- parts that are easy to dismantle and reuse at the next site.

No technical scale treatment plants for P-elimination at mining lakes have been built in Germany to date. The treatment of the inflow has been performed by using existing water bodies as pre-basins as described in the TGL 27885/02 (1983) standard. For filling the mining lake Cospuden, south of Leipzig, Germany, a slow sand filtration had been investigated on a pilot scale. The actual solution with untreated groundwater from the active opencast mine Profen has been described in Section 2.

8.3 In-lake Measures Against Eutrophication

Desludging is only a solution to eutrophication in very old mining lakes (#9 in Fig. 18). Even several decades after lake formation, the remaining mining structures can disturb all types of suction dredging. Only the furrows fill with soft sediments while elevated areas of the lake bottom are free from sediments.

The potential to eliminate nutrients with the help of macrophytes has been discussed (#10 in Fig. 18). Only a small share of the nutrient budget will be incorporated in the biomass and may be harvested by reed cutting. The intake is performed mainly in the early summer during the growth period. In the cold season the delivery of matter from degrading plants predominates.

Predator fish may be introduced as a top-down biomanipulation of the nutrient pyramid. With a subsequent reduction of prey fishes, the zooplankton have a better chance to survive and, therefore, to graze on phytoplankton. With less phytoplankton, the water becomes clear and more usable (#11 in Fig. 18). There are also some conflicts with fishery management on the mining lakes. One of the most important fish species to make use of the relatively scarce nutrient resources of the free water in the pelagic zone is *Coregonus albula*. Feeding mainly on zooplankton, the introduction of coregonids has a negative effect with respect to the control of eutrophication (Section 6).

Nutrient precipitation within the water body (#12 in Fig. 18) may be an appropriate option, if blue-green algae predominate due to a high phosphorus contents. Water blooms interfere with bathing activities. Dead algae have a stinking odour and cause the development of germs. Consequently bathing has to be prohibited.

At a 100-ha mining lake generated by gravel excavation near Magdeburg, Germany, a precipitation has been performed. A 490-t solution of aluminium sulphate was applied, corresponding to 5.7 g Al^{3+} per m^3 of lake water. Since this application the phosphorus concentration remains at a low level, the visibility of the Secchi disc is high (about 5 m during the summer time), and water blooms no longer occur. The surprisingly long lasting effect seems to be caused by sediment capping with aluminium hydroxide and aluminium phosphate, hindering the re-dissolution of the sediment phosphorus also under anoxic conditions (Klapper 1994). Before applying such an expensive chemotherapy to lakes from lignite mining, all further possibilities should be considered. Cheaper materials with P-binding properties may be found in the vicinity, such as ashes, clay overburden materials, iron-rich groundwater, etc.

Deep water aeration and destratification (#13 in Fig. 18) may be applied against eutrophication, if there is evidence of a better phosphorus binding due to a higher redox potential at the water-sediment interface. This effect has been overestimated in the past. Nevertheless, aeration or circulation of the deep water at least provides the benefit of a larger habitat area for fishes and their food supply. A combination of deep water aeration and the addition of iron-rich well waters has improved the water quality in the mining lake Kayna-South, Germany (Scharf 1998). The delivery of deep water (#14 in Fig. 18) is only an option for lakes with high through-flow. In a stratified lake, during stagnant periods, a vertical gradient is observed with high nutrient concentrations and low oxygen content in the deepest parts of the lake. This causes a trap effect and natural eutrophication. Similar to reservoirs with bottom outlets, deep water delivery in both natural lakes and mining lakes, with help of the Olszewski-tube, increases the export of nutrients and the export of oxygen deficit. The nutrient load decreases and the oxygen balance improves. If deep-water delivery is planned in the case of mining lakes, the necessary tubes may be easily installed in the empty basin before filling the lake.

The water level in the lake may be elevated with the aim of improving trophic conditions associated with a greater depth and volume (#15 in Fig. 18). The final water surface level of the mining lake should be at least in equilibrium with the adjacent groundwater table. To artificially keep down the lake level, e.g., to protect low lying buildings erected during the time of mining operation, should be avoided, if possible. Keeping a low lake level would be inconsistent with the principle of sustainability, as well as with the target of a good water quality with a low trophic level. At relatively shallow, older mining lakes, the harvesting of macrophytes is sometimes the only possible way to allow continued recreational and fisheries use (#16 in Fig. 18). During intensive macrophyte growth most of the phosphorus is incorporated in the plant biomass and may be eliminated with the harvested weed (Sampl 1979).

Another option may be to leave the plants as they are and to observe the succession towards a wetland. Distinct silting or landing stages are rare in "cultivated" country and may be interesting objects of nature protection, too.

The application of copper sulphate for eliminating blue-green algae should be taken into consideration only in extreme situations when it is necessary to guarantee the drinking water supply (#17 in Fig. 18). In the warmer countries of North and South America copper application at drinking water reservoirs is quite common, even though, because of the fish toxicity, only subtoxic amounts may be applied (Klapper 1998b). In bottom sediments copper is present mostly in a less toxic carbonate form, so that benthic animals, the food for fish, may generally develop in a normal fashion. Nevertheless, the chemical elimination of the primary producers in general should be used only as a last option. Moreover, in mining lakes there is the potential for very acidic groundwater, coming from the overburden heaps in the form of dilute sulfuric acid, to re-dissolve toxic copper compounds in the sediments.

Water quality management in connection with the control of eutrophication has been described in several books, which are recommended for more detailed information: Cooke et al. (1993); Ryding and Rast (1989); Sas (1989); Harper (1992); Klapper (1991); Klapper and Scharf, in Brauer, ed. (1997); Busch et al. (1989).

9 Contamination

9.1 Decision Criteria for Treatment

Opencast mining holes have been used by mining companies and by the coal processing industry for dumping of overburden, ashes from power plants, coal dust from briquette factories, Winkler ash after phenol adsorption, wastes from smoulderies, tar, etc. They have even been used for deposition of municipal garbage and byproducts from other industries. Provided that there were no negative impacts on the water quality in the generating mining lakes and on the groundwater in the surroundings, this policy follows an old mining law, to give back the greatest area possible for agricultural and forestry use. This refilling with rubble and waste is mainly used to reduce the deficit of material remaining from the abstraction of coal. Regarding the water quality management of the mining lakes and water protection in general, complicated investigations of the character of the dumped materials have to be performed.

Along with the most recent progress in analytical methods, there is new knowledge of ecotoxicology. However, the problems generated by the disposal of different materials into mining lakes are complex. Each substance behaves differently under different environmental conditions such as pH, Eh, temperature, conductivity, matrices of other dissolved substances, etc., and with respect to solubility, degradability or persistence, bioaccumulation and other characteristics.

Today hundreds of compounds may be analysed in effluents from various deposits. However, with such data alone, the assessment of ecotoxicological impacts is very difficult. Therefore, besides the chemical analyses, ecotoxicological effects should be investigated with help of an inventory of indicator organisms with respect to saprobity, trophy, halobity, etc. With the components of the biocenosis, a good diagnosis of ecological health and toxicological impacts is possible.

Liquids deposited in old mining sites are generally more critical than solid materials. Liquids may spread into the groundwater to an extent depending upon the chemistry (sorption, biodegradation) of the substances and the permeability of the subsurface material.

A special technology that has been applied in the past in the case of very toxic or concentrated liquid wastes is high pressure injection into deep geological layers, e.g. into plate dolomite. However, examples are known where such hazardous substances reappear at large distances nearly undegraded or undiluted at the surface. Therefore, this deep well injection has been prohibited in Germany. Nevertheless, the dangers generated decades ago persist and still have to be monitored at several locations.

The water of the mining lakes in eastern Germany is polluted with hazardous substances not only from dumping sites. During dewatering of the uranium mines in the Ronneburg district, the Weisse Elster river was critically loaded with radionuclides. Where sulfur-acidified groundwater is flowing into to a mining lake, relatively high contents of heavy metals have been observed, sometimes at toxic concentrations. This depends on the geological composition of the area.

In the case of saltcoal mining the NaCl-content is very high in water. Salt may be curbed only by hydraulic counterpressure and by dilution.

During the microbial degradation of hazardous organic substances new metabolites may occur. The most important question for the assessment of impacts is whether or not degradation takes place at all. The speed of biodegradation may be increased by ecotechnological measures.

For possible utilization of secondary raw-material, the chemical industry has sometimes dumped distinct byproducts separately. However, when no option for a second use was taken, such single-substrate deposits prove to be more difficult to degrade compared with municipal mixed dumps. The latter have less hazardous substrates, more adsorbing ashes, degradable substrates, stimulating co-metabolisms, etc. Compounds, being quasi conserved by substrate inhibition because of high concentrations need to be sufficiently diluted and supplied with nutrients, oxygen and other factors limiting biodegradation.

Deposits of different substances may affect the fisheries in mining lakes in different ways. Organic degradable substances interfere with the oxygen balance. Salt causes meromictic stable stratification and anoxic deep waters, killing the eggs of the bottom-breeding Coregonid fish.

Deposits deliver substances such as heavy metals to the mining lakes, which may accumulate at different rates in the various organs of the fish. Ammonium in high concentrations, such as in the deposit at the bottom of the mining lake Großkayna near Merseburg, Germany, excludes fishes from the lake. Lipophilic compounds in particular accumulate in fatty flesh of fish, impairing the taste. Fat fishes, like eel and carp, are particularly affected by lipophilic matter.

9.2 Approaches for In-situ Remediation and Ex-situ Treatment of Contamination at Mining Lakes

A systematic record of all deposits in the mining lakes of eastern Germany and the assessment of their environmental impact revealed a great number of different conditions. Therefore, a scheme was developed outlining different approaches for treating the deposits and contamination in and around mining lakes, with methods for solving the problems in an ecologically acceptable and economically feasible ways (Table 9). For all cases mentioned in Table 9 examples occur in the new federal provinces of Germany. Usually combinations of different chemical, physical and microbiological conditions have to be assessed, to prepare plans for adequate treatment.

At the Nachterstedt mining hole a waste dump has been removed. This was very expensive, but was necessary for construction of a planned recreational center for water sports.

The isolation of a deposit at the side of the mining lake Hufeisensee in Germany is the target at Halle-Kanena. A huge mixed municipal and industrial landfill occupies a most of an opencast hole. The other part of the hole has been filled with overburden and the rest with groundwater. Experts were afraid that the lake would be impaired by the inflow of contaminated groundwater passing through the deposit (Carmienke et al. 1987; Glässer and Klapper 1992; Christoph 1995; Gläser 1995). However, a virtually water-tight soil cover, due to vegetation, allows little infiltration of water through the dump site. Fortunately, a deep trench in the mining lake is located adjacent to the landfill. The water seeping from the deposit only reaches the trench and does not mix with the rest of the water in the lake. The small monimolimnion obviously functions as a trap for hazardous substances. Limnological investigations show a high diversity of sensitive species in the lake. Therefore, bathing is now licensed in some areas of the lake, despite the accumulation of hazardous matter buried deep at the bottom of the lake (Scharf et al. 1995).

Under favourable conditions the deposit may be isolated within the lake itself. In the deepest parts of the lake the isolation of dumped waste is possible by capping with inert materials. A well-documented capping with sand has been performed in Hamilton Harbour, in Lake Ontario, Canada, to separate toxic sludges from the Hamilton steelwork from the overlying water (Zeman and Patterson 1997). In mining lakes very often nature provides this isolation without additional human activities. Where precipitated iron hydroxide and plankton biomass are covering the deposit in sufficiently thick layers, the best solution for the dumping site is not to interfere with the natural processes. On the other hand, if wastes had been dumped around the edges of the mine, it has to be considered that waves and erosion might again expose the dump, even with thick layers of covering materials. A total subhydric incapsulation, including the lake surface, was the final solution of one of the most complicated deposits in a mining lake. The mining lake Silbersee in Germany was used for dumping different matter from the Wolfen film factory. From the original volume of $2.7 \times 10^6\,m^3$ today $1.1 \times 10^6\,m^3$ is occupied by spoiled ashes, $0.4 \times 10^6\,m^3$ by carbonate sludge from a water treatment plant and $1 \times 10^6\,m^3$ by lignine and cellulose sludge from a cellulose factory. The latter sludges are of a pudding-like consistency, with more than 90 % water content. A high content of sulfur led to gas emissions of sulfur hydrogen, mercaptane and other stinking and poisonous gases. Trials with an ex-situ pilot plant led to a solid-liquid separation. But no plants are growing on the solid material and the liquid phase did no receive license as an effluent.

Table 9. Outline of approaches to treat contaminated deposits at mining lakes

Removal of the waste dump	Isolation of the contamination near or in the lake		Restoration of the contamination			
			ex-situ treatment		in-situ treatment	
	general	*subhydric deposition*	*off site* in existing treatment plant (transport of matter is necessary)	*on site* mobile treatment plant	*partitioned* enclosures limnocorrals subhydric straw reactor	*lake sites* mixolimnion monimolimnion hypolimnion sediments
transfer of the waste to a special deposit	covering by plastic sheets or by soil with evaporating vegetation	profundal: capping with sand, ash or overburden; covering with matter from the pelagial: dead algae, Fe-hydroxide	physical methods sedimentation, flotation, stripping, adsorption, combustion		precipitation and transfer to the permanent sediment	
sorting, screening, deposition of only hazardous parts of the waste	technical capsulation by sheet piling or grout curtain		chemical methods precipitation, dissolution, ion-exchange, oxidation, extraction		neutralisation (liming) change in solubility abatement of H_2S with Fe^{3+}, NO_3- and liquid O_2	
drying; combustion	geohydraulic: inverted wells preventing groundwater influx	littoral: covering with soil, planting macro-phytes	biological methods mixing reactor, oxic: activated sludge (with P) solid bed reactor, oxic: trickling filter, contact aerator		aeration: surface aeration deep water aeration destratification	
utilization as a secondary resource						
pumping of liquid wastes (e.g. tar) to a treatment	lowering well in the dump preventing outflow from the deposit	water surface: covering with biofilters to prevent odour	solid bed reactor, anoxic: microbial desulfurication		incorporation in biomass; metabolisation; addition of limiting nutrients	
	treatment of the well water as sewage	covering with floating reed				

After all experiments had failed, the new target was the encapsulation of the material. Fortunately, groundwater contamination may be excluded in this case, because the fine-grained materials are nearly watertight. The deposit is like a sealed capsule in the regional groundwater field and the groundwater flows around it without causing significant pollution. Where the sludge reached the surface of the lake, 3-ha has been covered with biofilters – bags, filled with a mixture of wood sticks and foamed polystyrole chips (Fig. 19). The remaining approximately 1-m deep or shallower lake area is kept aerobic with the help of floating rotor aerators to avoid further H_2S emissions (Schmitz 1996). Desired spreading of floating reed, as in the mining lake Haselbach 1 in Germany, needs a surprisingly long time. This is probably due to the sulfur hydrogen, poisoning the cells of the rhizomes. The new target for this lake is to form a green covered area, protected against humans and reserved for nature and nature observation.

Nearly all ex-situ treatments, either off-site in existing treatment plants or on-site immediately at the deposit, are very expensive and generally not satisfactory with respect to the through flow. The off-site process is expensive because of the necessary transport of the polluted media to existing facilities, such as sewage treatment plants or incinerating plants.

For the on-site treatment special facilities have to be installed and operated for all necessary steps of physical, chemical and biological purification. These facilities are available as modules that may be combined according to the technological demand. For on-site operation, electricity and pipes have to be installed from the deposit to the treatment modules and from the treatment plant to polishing and accumulation ponds, e.g., for oxidation. Moreover, on-site operation includes human attendance during night and day. The process steps are mainly the same for on-site and off-site treatment plants and are selected according to the matter to be treated.

The most interesting developments are different in-situ treatments. These may be classified as part of the modern field of ecotechnology, i.e., biotechnology on the level of the ecosystem. With scientific knowledge of the natural functions within these ecosystems, these approaches aim only to help the system help itself by metabolisation and cometabolisation, incorporation, bioaccumulation, bioflocculation and sedimentation of the treated materials. All these processes may be identified as self-purification. To stimulate aerobic processes, oxygen has to be supplied. For anaerobic processes, such as microbial neutralisation by desulfurication, oxygen has to be kept away with help of hydraulic isolation or biochemical oxygen consumption.

The developing technologies are at the stage of pilot scale experiments in artificially separated parts of the water body such as enclosures, limnocorrals or in submerged reactors like the straw-cage for heterotrophic denitrification mentioned above. On a full, technical scale those lake sites with the most suitable environmental conditions have to be selected depending on the targets to be met. Clearly, sediments and the deep waters are suitable for anoxic treatments, while the surface waters or destratified or aerated parts of the water body are more suitable for oxic degradation.

For decades effluents from the lignite pyrolysis plants formerly operated in Central Germany have contaminated the region's groundwater and surface water. In order to develop remediation strategies, the case study of a pyrolysis waste water deposit was used to investigate the complex characterisation of the deposit as well as the fundamental chemical and biological reactions of pyrolysis waste water altered by storage. The group of anthropogenic humic substances (AHS), formed by abiotic oxidative polymerisation from monomeric phenols, was perceived as determining the chemical behaviour and the biological effects (Stottmeister et al.1998).

The deposit in a former open cast mine, located in Saxony-Anhalt, near the city of Zeitz (village Trebnitz), showed the following characteristics when the studies began in 1992:

- area 9 ha,
- water volume 2 million m^3,
- maximum depth 27 m,
- visibility depth 0.03 m,
- oxygen content of the water 0 mg l^{-1}.

A strong odour of organic sulfur compounds irritated the inhabitants of the region.

The dark brown water body showed a stratification with concentration of total phenolic compounds between 7 to about 100 mg l^{-1} in the upper zone (0–10 m) and of more than 200 mg l^{-1} in the deeper ones (15–24 m). Ammonia was detected in the same concentration range (up to 250 mg l^{-1}). The maximum of the COD (chemical oxygen demand) was 2,290 mg l^{-1}, the maximum DOC (dissolved organic carbon) 690 m l^{-1}in 20 m depth.

A traditional ex situ water treatment was not practicable. A new unconventional in situ remediation strategy was developed and tested in *in situ* enclosures (see Fig. 20 and 21) (Stottmeister et al. 1999). The positive results of these experiments were used for a full scale remediation of the deposit (Stottmeister et al. 2000).

The strategy can be described very shortly as follows: the initiation of the biological self purification in a stable stratified water body with an aerobic upper and an anaerobic deeper zone and a sediment area (application of the "pontos euxinos" effect).

In detail, the following steps of "habitat engineering" were carried out:

- flocculation of black AHS-macromolecules by iron-(III)-chloride solution (3200 m^3 40%) at pH 4–5, resulting in a sedimentation of the flocs and settling on the ground;
- neutralisation of the formed clear uncoloured water body with 50 % of the former DOC (pH 7–8) with limestone suspension (2,200 m^3 20%);

- addition of phosphoric acid in steps (3 x 0.8 m³ 75 %).

The most sensitive step was the injection of the chemicals into the water body because the specific water density had to be calculated for a stable stratification. Like in the enclosure experiments, after flocculation of the toxic and oxygen consuming AHS macromolecules the clear water was immediately saturated with oxygen down to 3–4 m (depending on the season) and the microbiological degradation activity started after neutralisation. The addition of phosphorus (the limitation of P was caused by the addition of Fe(III)-salt) initiated an intensive algae growth with photochemical oxygen production and additional sorption capacity for residues of AHS.

In only three years (1998–2001) both the biotic and abiotic processes together resulted in the formation of a non-toxic ecosystem with a "normal" hydrobiological biodiversity in the upper aerobic zone (down to 6 m). No phenols are detectable in the surface zone (down to 5 m). The phenols are eliminated also to 95 % in the depth of 20 m indicating anaerobic (methanogenic and iron-III-reducing) activities.

The long-term behaviour of this first ever realised "enhanced natural attenuation"-process of an industrial waste water deposit is under studies.

Fig. 19. Deposit of chemical wastes in a former opencast mine at Wolfen, Germany. The surface is covered with biological filters protecting against H_2S emissions (photo: U. Stottmeister)

Fig. 20. Installation of an enclosure for sediment investigation in the Schwelvollert deposit in Saxony-Anhalt, Germany (field studies of the UFZ Centre for Environmental Research, Leipzig, photo: (U. Stottmeister)

Fig. 21. Field laboratory with several enclosures in a deposit of wastewaters from lignite carbonization (UFZ Centre for Environmental Research, Leipzig, photo: U. Stottmeister)

References

Babenzien HD (1996) Studien zur Mikrobiologie eines sauren Gewässers (Experimentalgewässer Gr. Fuchskuhle). Jahrestagung DGL Schwedt, Abstracts

Bell PE, Mills AL (1987) Appl Environ Microbiol 53:2610–2616

Bell RT, Vrede K, Stensdotter-Blomberg U, Blomquist P (1993) Stimulation of the microbial food web in an oligotrophic, slightly acidified lake. Limnol Oceanogr 38(7):1532–1538

BLAG (Bund-Länder-Arbeitsgruppe Wasserwirtschaftliche Planung) (1994) Rahmenkonzept zur Wiederherstellung eines ausgeglichenen Wasserhaushaltes in den vom Braunkohlenbergbau beeinträchtigten Flußeinzugsgebieten in der Lausitz und in Mitteldeutschland. Beschluß vom 17/18.3.94 (unpublished)

Brauer H (ed) (1997) Handbuch des Umweltschutzes und der Umweltschutztechnik. B.5 Sanierender Umweltschutz. Springer Berlin Heidelberg New York, 410 p

Brettschneider U, Pöpel HJ (1992) Sulfatentfernung aus Wasser. Wasser-Abwasser-Praxis 1: 298–304

Brown RL (1971) Removal of nitrogen from tile drainage – a summary report. Agricult. Wastewater Studies Rep. No. 13030 ELY 7/11. DWR Bulletin:174–179

Busch K-F, Uhlmann D, Weise G (eds) (1989) Ingenieurökologie. Fischer Jena 488 p

Carmienke I, Tauchnitz J, Kiesel G (1987) Mikrobiologische Untersuchungen an deponiebeeinflußten Wässern unter besonderer Berücksichtigung sulfatreduzierender Bakterien. Acta hydrochim hydrobiol 15:587–602

Christoph G (1995) Modellrechnungen zur Entwicklung des Hufeisensees in der Nähe der subhydrischen Deponie Halle-Kanena. UFZ report Nr 4/95, 81–83

Cooke GD, Welch EB, Petersen SA, Nawroth PR (1993) Restoration and management of lakes and reservoirs. 2nd ed. Lewis Publishers Boca Raton, AnnArbor, London, Tokyo 558 pp

Davison W, George DG, Edwards NJA (1995) Controlled reversal of lake acidification by treatment with phosphate fertilizer. Nature 377:504–507

Dietz JM, Watts RG, Stidinger DM (1994) Evaluation of acidic mine drainage treatment in constructed wetland systems. 3rd Intern. Conf. on the Abatement of Acid Drainage, vol 1: Mine Drainage. U.S. Bureau of Mines Special Publication SP 06 A-94, 70–79

Fichtner N (1983) Verfahren zur Nitrateliminierung im Gewässer. Acta hydrochim hydrobiol 11:339–345

Friedrich H-J (1996) Ausarbeitung eines Verfahrens zur Sanierung saurer Tagebaurestseen durch elektrolytische Wasserstoffabscheidung und Schwermetallfällung. 1. Statusseminar zur BMBF-Fördermaßnahme „Sanierung und ökologische Gestaltung der Landschaften des Braunkohlenbergbaues in den neuen Bundesländern“ Cottbus, proceedings 71–72

Friese K, Wendt-Potthoff K, Zachmann DW, Fauville A, Mayer B, Veizer J (1998) Biogeochemistry of iron and sulfur in sediments of an acidic mining lake in Lusatia, Germany. Water, Air and Soil Pollution

Geller W, Klapper, Salomons W (eds) (1998) Acidic mining lakes. Acid mine drainage, limnology and reclamation. Springer, Berlin Heidelberg New York 435 pp

George DG, Davison W (1998) Managing the pH of an acid lake by adding phosphate fertilizer. In: Geller W, Klapper H, Salomons W (eds) Acidic mining lakes. Springer, Berlin Heidelberg New York, 365–384

Gewässergütebericht Sachsen-Anhalt (1992) Landesamt f. Umweltschutz Sachsen-Anhalt (ed) Halle 1993. p 399

Gläser HR (1995) Geophysikalische Untersuchungen zum Nachweis von Fließwegen –

Fallstudie Hufeisensee. UFZ report Nr 4/95, 94–102

Glässer W, Klapper H (1992) Stoffflüsse beim Füllprozeß von Bergbaurestseen – Entscheidungsvorbereitung für die Sanierung von Tagebaulandschaften. Boden, Wasser und Luft. Umweltvorsorge in der AGF-Arbeitsgemeinschaft der Großforschung, Bonn, 19–23

Glombitza F, Madai E (1996) Entwicklung und Erprobung eines Verfahrens zur Behandlung saurer, sulfatreicher eisenhaltiger Wässer aus dem Braunkohlenbergbau. 1. Statusseminar zur BMBF Fördermaßnahme „Sanierung und ökologische Gestaltung der Landschaften des Braunkohlenbergbaues in den neuen Bundesländern. Cottbus, proceedings 68–70

Harper D (1992) Eutrophication of freshwaters. Principles, problems and restoration. Chapman and Hall London, 327 pp

Hedin RS (1989) Treatment of coal mine drainage with constructed wetlands. In: Majumbar SK et al. (eds) Wetland. Ecology and Conservation: Emphasis in Pennsylvania. The Pennsylvanian Academy of Science, Pennsylvania, pp 349–362

Hedin RS, Watzlaf GR (1994) The effect of anoxic limestone drains on mine water chemistry. 3rd Intern Conf on the Abatement of Acid Drainage. 1: Mine Drainage. U.S. Bureau of Mines Special Publication SP 06 A-94, 185–194

Jones JR (1974) Denitrification by anaerobic filters and ponds – phase II. EPA Rep. No. 13030 ELY 06/71–74

Kepler DA, McCleary EC (1994) Successive alkalinity-producing systems (SAPS) for the treatment of acid mine drainage. U.S. Bureau of Mines Special Publication SP 06 A-94, 195–204

Klapper H (1991) Control of eutrophication in inland waters. Ellis Horwood, Chichester 337 pp

Klapper H (1994) Technologie zur Verbesserung der Wasserbeschaffenheit eutrophiertes Seen. Angew Gewässerökol Norddeutschlands, H.7:40–45

Klapper H (1995) Bergbau-Restseen – Wassergüteprobleme. Vortragsband des 1. GBL-Kolloquiums März 1995 Leipzig. E. Schweizerbart´sche Verlagsbuchhandlung Stuttgart, pp 20–35

Klapper H (1998a) Sanierungsstrategien für die Braunkohlegewässer unter besonderer Berücksichtigung der Anforderungen an die Wasserqualität für eine fischereiliche Nutzung. Studie für das Inst. f. Binnenfischerei Potsdam-Sacrow, p 57 (unpublished)

Klapper H (1998b) Water quality problems in reservoirs of Rio de Janeiro, Minas Gerais and Sao Paulo. 3rd Int. Conf. on Reservoir Limnology and Water Quality, proceedings Budweis 1997. Internat Rev Hydrobiol 83 Special Issue:93–102

Klapper H (ed) (1980) Flüsse und Seen der Erde. URANIA-Verl. Leipzig, 240 p

Klapper H, Friese K, Scharf B, Schimmele M, Schultze M (1998) Ways of controlling acid by ecotechnology. In: Geller W, Klapper H, Salomons W (eds) Acidic mining lakes. Acid mine drainage, limnology and reclamation. Springer, Berlin Heidelberg New York, 401–416

Klapper H, Geller W, Schultze M (1996) Abatement of acidification in mining lakes in Germany. Lakes and Reservoirs: Research and Management, No 2:7–16

Klapper H, Scharf B (1997) Sanierungstechnologien, Standgewässer. In: Brauer H (ed) Handbuch des Umweltschutzes 5: Sanierender Umweltschutz. Springer Berlin Heidelberg New York, 172–223

Klapper H, Schultze M (1993) Das Füllen von Braunkohlerestseen. Wasserwirtschaft Wassertechnik H.5:34–38

Klapper H, Schultze M (1995) Geogenically acidified mining lakes – living conditions and possibilities of restoration. Int Revue ges Hydrobiol 80:639–653

Klapper H, Schultze M (1996) Geogene Versauerung von Bergbaurestseen: Ursachen,

Folgen, Möglichkeiten zur Schadensbegrenzung. GBL-Gemeinschafts-vorhaben H. 3, E. Schweizerbart'sche Verlagsbuchhandlung Stuttgart, pp 121–131

Klapper H, Schultze M (1997) Sulfur acidic mining lakes in Germany – ways of controlling geogenic acidification. IVth Int. Conf. on Acid Rock Drainage Vancouver. proceedings IV, 1727–1744

Kleinmann RLP, Crerar DA, Pacilli RR (1981) Biogeochemistry of acid mine drainage and a method to control acid formation. Mining Eng:33, 300–304

Kringel R (1998) Untersuchungen zur Verminderung der Auswirkung von Pyrit-oxydation in Sedimenten des Rheinischen Braunkohlenreviers auf die Chemie des Grundwassers. Diss. Ruhr Univ. Bochum, Diss. Druck Darmstadt, p 176

Kringel R, Schultze M, Wölfl S, Büttner O, Kellner S, Herzsprung P (1997) Limnologisches Gutachten für den Tagebaukomplex Goitsche – Teil 2. UFZ-Umweltforschungszentrum Leipzig-Halle GmbH, Sektion Gewässerforschung Magdeburg (unpublished)

Lamb HM, Dodds-Smith M, Gusek J (1998) Development of a long-term strategy for the treatment of acid mine drainage at Wheal Jane. In: Geller W, Klapper H, Salomons W (eds) Acidic Mining Lakes, Springer Berlin Heidelberg New York, 335–346

Luckner L, Eichhorn D, Gockel G, Seidel K-H (1995) Durchführbarkeitsstudie zur Rehabilitation des Wasserhaushaltes der Niederlausitz auf der Grundlage vorhandener Lösungsansätze. Herausg. Dresdner Grundwasserforschungszentrum e.V., 90 p

Lüderitz V (1988) Kupferwirkung auf Planktonalgen. Dissertation A, Humboldt Univ. Berlin, Math.-Nat. Fak.

Nawroth JR, Conley PS, Sandusky JE (1994) Concentrated alkaline recharge pools for acid seep abatement: principles, design, construction, and performance. 3rd Int. Conf. on the Abatement of Acid Drainage. U.S. Bureau of Mines Special Publication SP 06 A-94, 382–391

Nixdorf B, Fyson A, Schöpke R (1997) Versauerung von Tagebauseen in der Lausitz – Trends und Möglichkeiten der Beeinflussung oder: Kann die biogene Alkalinitätsproduktion gesteuert werden? DGL Tagungsbericht 1996 (Schwedt) 513–517

Nixdorf B, Rücker J, Köcher B, Deneke R (1995) Erste Ergebnisse zur Limnologie von Tagebaurestseen in Brandenburg unter besonderer Berücksichtigung der Besiedlung im Pelagial. Limnologie aktuell 7:38–52, G. Fischer Verlag

Nolte E (1947) Reinigungsanlagen für Phenolabwässer nach dem Magdeburger P-Verfahren. Wasser-, Abwasser- u Fischereichem 2:3–14

Ohle W (1981) Photosynthesis and chemistry of an extremely acid bathing pond in Germany. Verh Int Ver Limnol 21:1172–1177

Oksijuk OP, Stolberg FW (1986) Control of water quality in canals (in Russian) Naukova Dumka Kiev, 173 pp

Olem H (1991) Liming acidic surface waters. Lewis publishers, 331 pp

Olofsson H, Blomquist P, Olsson H, Broberg O (1988) Restoration of the pelagic food web in acidified and limed lakes by gentle fertilization. Limnologica 19:27–35

Philips P, Bender J (1998) Bioremediation of metals in acid tailings by mixed microbial mats. In: Geller W, Klapper H, Salomons W (eds) Acidic mining lakes. Springer Berlin Heidelberg New York, 347–363

Pietsch W (1993) Gewässertypen der Bergbaugebiete des Lausitzer Braunkohlereviers und Makrophytenverbreitung. In: Abschlußbericht zum Forschungsprojekt „Erarbeitung der wissenschaftlichen Grundlagen für die Gestaltung, Flutung und Wassergütebewirtschaftung von Bergbaurestseen" BMFT-Förderkennzeichen 0339450 B

Rastogi V (1996) Water quality and reclamation management in mining using bactericides. Preprint from Mining Engineers, 6 pp

Reichel F, Uhlmann W (1995) Wasserbeschaffenheit in Tagebaurestseen. Studien und Tagungsberichte 6, Landesumweltamt Brandenburg: p 86

Ryding S-O, Rast W (eds) (1989) The control of eutrophication of lakes and reservoirs. Man and the biosphere series 1, UNESCO and Parthenon Publ. Group Paris, 314 pp

Sampl H (1979) Die Massenentwicklung der Wasserpest *Elodea canadensis* im Afritzer See im Jahre 1978. Kärntner Naturschutzblätter 17:59–70

Sas H (ed) (1989) Lake restoration by reduction of nutrient loading. Expectations, experiences, extrapolations. Academia Verl. Richarz, Sankt Augustin, 497 pp

Scharf BW (1998) Abschlußbericht für den Tagebaurestsee Kayna-Süd. UFZ – Sektion Gewässerforschung Magdeburg. 9 p 6 Anl (unpublished)

Scharf BW, Dermietzel J, Melzer A, Richter W, Rönicke H (1995) Beiträge zur Limnologie des Hufeisensees bei Halle/Saale. UFZ report Nr. 4/95, 53–80

Schindler DW (1994) Changes caused by acidification to the biodiversity: Productivity and biochemical cycles of lakes. In: Steinberg C., Wright J (eds) Acidification of freshwater ecosystems: Implications for the future. Wiley and Sons, New York, pp 154–164

Schmitz HJ (1996) Schwimmender, begehbarer Biofilter als großflächige Sicherungsmaßnahme auf dem Silbersee (Grube Johannes). TerraTech, Zeitschrift für Altlasten und Bodenschutz 5:49–52

Stottmeister U, Glässer W, Klapper H, Weissbrodt E, Eccarius B, Kennedy C, Schultze M, Wendt-Potthoff K, Frömmichen R, Schreck P, Strauch G (1999) Strategies for Remediation of Former Opencast Mining Areas in Eastern Germany. In: Azcue JM (ed) Environmental Impacts of Mining Activities. Springer, Berlin Heidelberg New York, 3–296

Stottmeister U, Weißbrodt E, Becker PM, Pörschmann J, Kopinke FD, Martius GGM, Wießner A, Kennedy C (1998) Analysis, behaviour and fate of a lignite pyrolysis wastewater deposit. Contaminated Soil'98, Thomas Telford, London, pp 113–121

Stottmeister U, Weissbrodt, E (2000) Enhanced Bioattenuation. Anwendung natürlicher Prozesse zur Sanierung carbochemischer Altlasten. TerraTech, Zeitschrift für Altlasten und Bodenschutz 1:45–48

Sword BR (1971) Denitrification by anaerobic filters and ponds. EPA. Rep. No. 13030 ELY 04/71-8, 1–68

TGL 27885/01 (1982) Nutzung und Schutz der Gewässer/Stehende Binnengewässer/ Klassifizierung. Standard

TGL 27885/02 (1983) Nutzung und Schutz der Gewässer/Stehende Binnengewässer/Nährstoffelimination in Vorsperren. Standard

Ulrich B (1981) Die Rolle der Wälder für die Wassergüte unter dem Einfluß des sauren Regens. Agrarspektrum 2:212–231

Wendt-Potthoff K, Neu TR (1998) Microbial processes for potential in situ remediation of acidic lakes. In: Geller W, Klapper H, Salomons W (eds) Acidic mining lakes. Acid mine drainage, limnology and reclamation. Springer-Verlag Berlin, Heidelberg, New York, pp 269–284

Zeh E (1998) Seenlandschaft bereits in Sicht. Flutungskonzept für den Südraum von Leipzig. Informationsmaterial von MIBRAG, STRABAG und LMBV

Zeman AJ, Patterson TS (1997) Preliminary results of demonstration capping project in Hamilton Harbour. Water Qual Res J Canada 32:439–452

Ziegenhardt W (1994) Das Programm "Altlastensanierung Braunkohle" insbesondere seine Ziele und Aufgaben der wasserhaushaltlichen Sanierung. VDI-Berichte Nr. 119:87–100

Ziegenhardt W, Trogisch R (1996) Bedeutung des Speichersystems Lohsa II für die wasserhaushaltliche Sanierung im Spreegebiet. Grundwasser-Zeitschrift der Fachsektion Hydrogeologie 3–4:142–147

Development Prospects for the Post-Mining Landscape in Central Germany

Sabine Linke, Lutz Schiffer

Sächsische Akademie der Wissenschaften zu Leipzig – Saxon Academy of Sciences, Leipzig, Arbeitsstelle Technikfolgenabschätzung – Division for Technology Assessment, D-09107 Chemnitz, Germany

1 Introduction

The central German coalfield has an area of around 50,000 hectares, making it the third largest in Germany. It is regarded as the "classic location of German lignite mining" (Meyer et al. 1995) and contains about 33 % of the minable lignite stocks in eastern Germany.

Lignite mining dominated the economic and social development of Central Germany for centuries. Once uncompromisingly encouraged as a guarantor of economic prosperity and extensively pursued especially in the 20th century until the late 1980s as part of East Germany's drive for independence from raw material imports, it was responsible for ravaging the natural balance. The enormous changes caused to the landscape by lignite mining represent a massive intervention on a scale which in terms of its short duration is only comparable with individual catastrophes in the earth's history (Eissmann and Campen 2000). They encompass irreversible losses of countryside and severe ecological damage. Moreover, environmental pollution was exacerbated by the nearby coal-processing industry. The result for the local population was the constant depreciation of the conditions of everyday life, prompting above-average migration from Central Germany.

Both the mines and the coal beneficiation plants largely collapsed in the wake of the economic changes in the early 1990s, which precipitated a drastic drop in the demand for lignite. Of the 21 opencast mines run in 1989, only three are still in operation. Efforts to remediate disused opencast mines were intensified after 1990 in order to overcome the damage to the countryside and the rest of the environment caused by mining – a process which is essential in order to help the local economy flourish and to overcome the widespread social problems.

The difficult legacy connected with the area's mining history also harbours the chance to develop Central Germany in a manner geared towards the objective of sustainability. The region provides almost limitless scope for moulding as the landscape and its diverse functions are in a largely disrupted or even ruined state and can by means of remediation schemes be geared to a new spectrum of usage. The scope and complexity of the necessary recultivation measures and the related potential as well as the example nature of problem solutions for comparable regions raise the usage-orientated sustainable moulding of the local mining landscape into a process of European importance.

2 The Central German Coalfield

2.1 A Region in the Grip of Transformation

The central German lignite coalfield is concentrated on an area bounded by the towns of Gräfenhainichen, Kitzscher, Altenburg, Zeitz, Weissenfels, Mücheln, Merseburg, Halle, Bitterfeld and Gräfenhainichen. It comprises four core coalfields:

- Leipzig's southern region,
- Zeitz–Weissenfels–Hohenmölsen,
- Geiseltal,
- Gräfenhainichen–Bitterfeld–Delitzsch.

It also includes the two peripheral smaller coalfields at Halle (Merseburg East mine) and Röblingen (Amsdorf mine, west of Halle). In administrative terms, the central German coalfield is situated in the German states of Saxony, Saxony-Anhalt and Thuringia.

The extraction of lignite in opencast mines dates back 300 years, with the last 100 years in particular being characterised by highly extensive expansion. The comparatively simple mining conditions of the extensive deposits of coal were a major factor behind the region's rapid economic development as of the late 19th century. Partly due to the short transport routes and the existing water resources, coal beneficiation plants were established in the immediate vicinity of the mines. The increasing needs of the arms industry during World War II and later East Germany's efforts to achieve economic independence led to a further increase in lignite mining (cf. Thomae 1996). The three sectors of lignite mining and briquette production, coal chemistry and energy supply continued to dominate until the late 1980s. For example in 1989 some 70 % of the primary energy demand and 80 % of East Germany's electricity demand was covered by the usage of indigenous lignite (cf. BMU 1994). In connection with the political changes in 1989, the demand for lignite sharply declined. By 1998, the amount of lignite mined in the central German coalfield had been reduced by 87 % (cf. DEBRIV 1999). This led to the closure of most opencast mines. In two core coalfields (Geiseltal, Gräfenhainichen–Bitterfeld–Delitzsch) and a smaller coalfield (Halle), lignite mining was completely abandoned. In the remaining coalfields, merely three lignite mines are still in operation (see Fig. 1).

In Central Germany, the scope of land used for mining was around 470 km² in 1989 (cf. Berkner 1995), climbing to about 510 km² in 1994 (cf. Berkner 1998). In this connection, a total of some 30–35 billion m^3 of earth had been excavated from depths down to 135 m (cf. Eissmann and Campen 2000). These extensive measures were accompanied by long-term damage and destruction to the landscape and the natural environment. Thus it was that the expansion of opencast mines resulted in the loss of entire areas. For example, it led to the complete or at least partial excavation of about 120 built-up areas and the resettlement of over 47,000 inhabitants (cf. Berkner 1995). In addition, the necessary reduction of the water table over a broad area totally altered the regional water balance, which in turn also changed the character of naturally evolved sections of the landscape.

By the early 1990s, the share of mining land in the central German coalfield which had been made reusable amounted to 47.4 % (ibid.). Thanks to the subsequent intensification of remediation efforts, by 1998 this proportion had risen to about 56 % (cf. DEBRIV 1999). The extent of land still in need of recultivation is thus considerable (see Fig. 2).

Remediating the mining areas in Central Germany (including the mining pits) is all part of removing the legacy of lignite mining in eastern Germany. Given the extent of the tasks to be overcome, the necessary timescale and the funding required, this scheme can be regarded as Germany's most important environmental project (cf. Kabisch and Linke 2000). In addition to bringing about and maintaining public safety, the necessary remediation work will result in a normal water balance requiring virtually no aftercare and allow the diverse usage of the land once employed for mining (see Fig. 3). Carrying out the measures required will take a number of decades. Owing to the difficulties of assessing the areas to be rehabilitated, the costs are estimated at DM 14–16 billion (cf. Stuba 1999) – a figure which does not include making up for inadequate recultivation on areas released from mining surveillance prior to 1990.

The flooding of mining pits and reforestation schemes planned during the remediation of opencast mines mean that Central Germany is about to undergo a unique process of landscape renewal. In addition, moulding the mining landscape is part of the region's structural transformation. Central Germany is regarded as one of the potential growth regions – as was stated by the European Union at the "Future Conference for Central Germany" in July 1999 in Leipzig. The traditional areas of lignite mining, the chemical industry and power generation constitutes an important factor for its further economic development, since lignite remains the most important indigenous fuel source.

Fig. 1. Mining activities at United Schleenhain, 1999

Fig. 2. Internal dump area in the disused Espenhain mine south of Leipzig, 1998

In 1998 LMBV (the administrative company in charge of the mines in Lusatia and Central Germany) mined 6.4 million tonnes of raw lignite, which was mostly used to generate electricity in the region's base-load power stations (cf. SMWA 1999). The band-new lignite-fired Lippendorf Power Station is one of the most advanced power plants of this type anywhere in the world.

The shutdown of opencast mines and the closure of many lignite-processing companies they supplied caused the almost 60,000 jobs in mid-1989 to be slashed to around 3,000 by the end of 2000 (cf. Mibrag 1994; www.braunkohle.de 2001). To overcome the problems on the labour market and to keep the population stable, it is essential that new, sustainable industrial sectors take root here. The fundamental change from a region dominated by mining into an attractive lake district is an important condition for this. The development of the enhanced landscape for tourism and the subsequent expansion and strengthening of the endogenous economic potential contains one of the great challenges of the future. A successful, comprehensive regional landscape and structural change will also help alter the area's image and encourage the emergence of a new regional identity.

Fig. 3. Remediation of the 'spoilbank dunes' at the shutdown Espenhain mine, 1997 (the new Lippendorf Power Station can be seen in the background)

2.2 Restoring the Water Balance

Lignite mining has fundamentally changed the water balance in Central Germany – both quantitatively and qualitatively. The main problem comprises the dramatic reduction of the water table and the resulting shortage of groundwater (cf. Woschni et al. 2000). Lowering the depth of the groundwater over a total area of 1,100 km^2 by sometimes more than 100 m, enormous losses of non-renewable groundwater already detected and still to come on a scale of 10 billion m^3, and the relocation or capping of rivers and streams are just three of the complications caused. The loss of considerable sections of floodplains of the Mulde, White Elster and Pleisse has reduced the floodwater retention areas and impaired biotope structures dependent on the groundwater. Moreover, the groundwater suffered extraordinarily high pollution caused by the lignite industry. It will take 50–90 years until the state of the groundwater and surface water returns to normal (cf. Umwelt 2001).

The central German coalfield covers various hydrological areas. The mining district south of Leipzig, including the disused Merseburg-East opencast mine, is in the catchment area of the White Elster. As at least five aquifers are assumed to exist here, its hydrogeological situation is relatively complicated. The related mining areas north of Leipzig are influenced by the Mulde catchment area. Here there are two main aquifers, making the area simpler in hydrogeological terms. The western part of the central German coalfield with local lignite deposits in Geiseltal and Röblingen (Amsdorf) is in the catchment area of the Saale, which in turn comprises a series of hydrogeologically separate areas each with their own water balance (cf. LMBV 1999; Meyer et al. 1995).

Each of these areas has its own specific problems regarding the flooding of mining pits and the limnological quality of the lakes forming therein. These problems stem from the way in which flooding is carried out. There are basically two ways of doing so and compensating for the shortage of groundwater: allowing the groundwater to rise naturally, and external flooding with river or drainage water. These two methods differ greatly concerning the speed and hence the duration of flooding. Lakes take about 38–95 years to form with natural flooding – a period which can be cut to just 3–15 years in the case of external flooding (cf. Lehmann 1997; LMBV 1997). In fact, both methods tend to be applied together in order to ensure favourable conditions for the planned usage of these mine lakes once flooding is complete.

Both natural flooding and external flooding using river water may be prone to difficulties. Solely allowing the groundwater to rise (an approach which has only rarely been used in Central Germany; see Fig. 4) may result in the acidification of the lake, and the water in these lakes will probably only become neutral after some 10–15 years. However, when using river water, mention should be made of the problem of maintaining an essential minimum (cf.

Woschni et al. 2000) to ensure that the rivers concerned and the life they contain are not impaired. This type of external flooding is the main approach used for the mining pits in the northern and western parts of the central German coalfield now undergoing renaturalisation (see Fig. 5). It is especially advantageous if drainage from active opencast mines is available for external flooding – as is indeed the case in the area south of Leipzig. Here Article 13 of the EU Water Framework Directive is observed, which expressly calls for the usage of groundwater pumped out of opencast mines prior to coal extraction (cf. Kabisch and Linke 1999). Using water from active mining enables high-quality water to be discharged into mining pits undergoing flooding. This means that the local water resources can be used, minimising extraction from nearby rivers (whose water is anyway often polluted). Moreover, this type of flooding is relatively fast and contributes to the stability of the pit slopes.

In connection with the flooding of mine lakes in Central Germany, additional problems may also be caused by geological conditions and environmental pollution. For example in the areas north Leipzig, especially in the Bitterfeld district, as well as south of Leipzig, severe soil contamination has been encountered. This was caused by harmful emissions from the now shut-down carbo-chemical plants and briquette factories. By contrast, the area west of Leipzig is partly beset by the problem of rising saline water making for salty lakes as well as that of contaminated industrial sites, e.g. the usage of flushing dumps and the discharge of wastewater into the lake area (cf. Grün 1998).

Fig. 4. Groundwater rising naturally as remediation takes place in Restloch Südkippe, Borna-East (Leipzig's southern region), 2000

Fig. 5. Flooding system for diversion of surface water from the River Mulde into the future Goitsche mine lake north of Leipzig, 2000

Other risks in connection with the flooding and subsequent usage of mining pits result from the presence of numerous landfills in the area. Especially problematic are unsorted municipal landfills and small pits where in the past industrial waste (including hazardous waste) was simply dumped. As the groundwater rises, such areas partly come in contact with groundwater circulation. Although minimising the dangers caused thereby is imperative, it is also expensive and technology-intensive. Furthermore, hazard potentials can never be ruled out *a priori* – not even in the case of new, sorted landfills (see Fig. 6).

As a result of the extensive remediation of opencast mines, which included efforts to normalise the local water balance, 41 new lakes are expected to emerge in the former mining landscape of Central Germany by about the mid-21st century. These lakes will have a total area of about 150 km^2 and a volume of nearly 3 billion m^3 (cf. Table 1).

Table 1. Size of new mine lakes in parts of the central German coalfield

District	Surface area of new lakes (km^2)	Volume of new lakes (million m^3)
Gräfenhainichen–Bitterfeld–Delitzsch	42	551
Leipzig's southern region	51	1,140
Zeitz–Weissenfels–Hohenmölsen	18	475
Geiseltal	23	492
Halle (Merseburg East)	7	80
Röblingen (Amsdorf)	4	58
Central Germany (total)	145	2,796

Source: Remediation framework plans, LMBV 1999, own calculations

The largest of these mining pit lakes will after the completion of flooding in 2002 and 2008 be Goitsche north of Leipzig (13 km^2, 213 million m^3; see Fig. 7) and Geiseltalsee west of Leipzig (18.4 km^2, 410 million m^3; cf. Schultze and Klapper 1997). They will be among the 20 largest lakes in Germany (cf. Dachverein Mitteldeutsche Straße der Braunkohle e.V. 1999). It can be assumed that the new mine lakes will form new types of lake landscape to the west, north and south of the Leipzig-Halle agglomeration together with the existing natural and artificial lakes (cf. Fig. 8).

Fig. 6. Cröbern landfill, which was built on a dump at Espenhain mine, 2001

Fig. 7. Pier and water level gauge at Goitsche prior to flooding, 2000

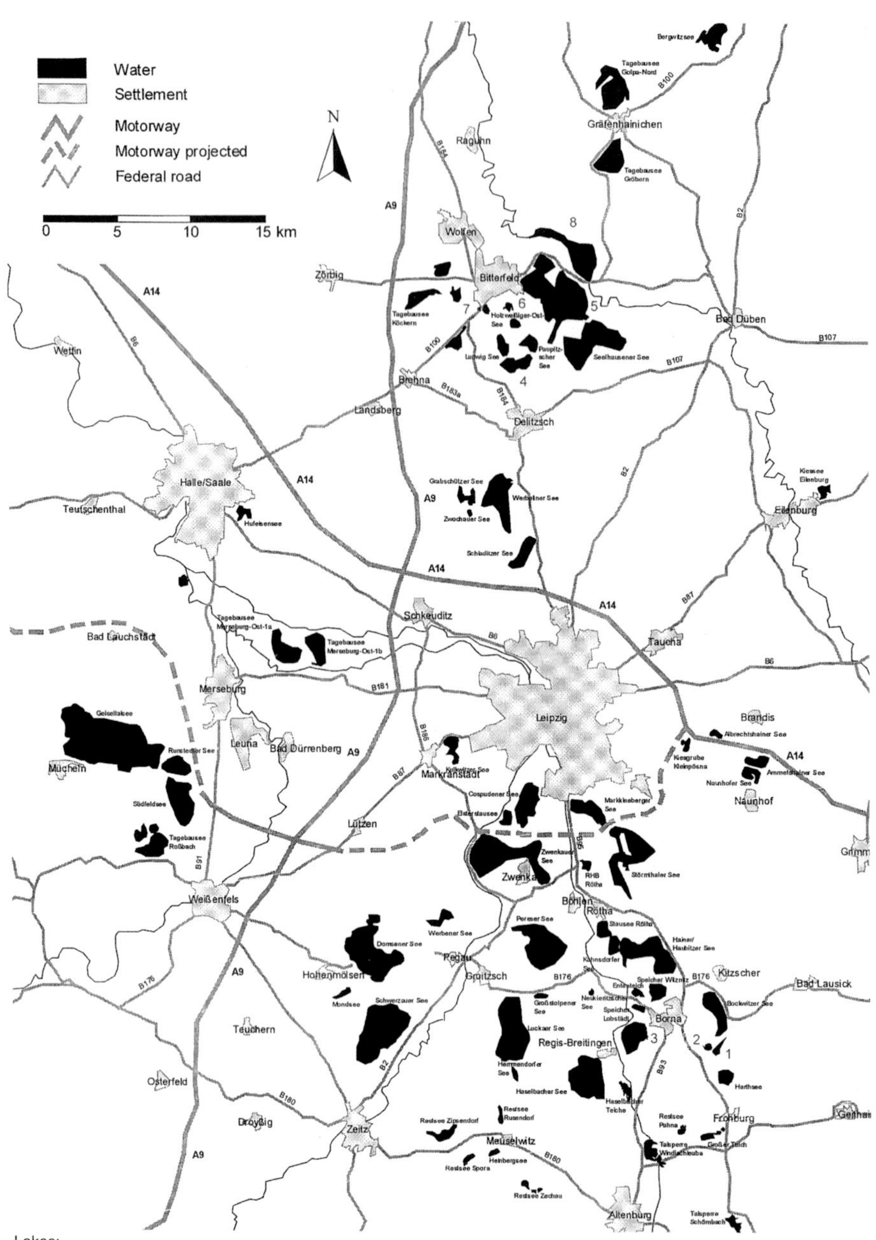

Lakes:

1 Südkippe, 2 Hauptwasserhaltung, 3 Speicher Borna, 4 Neuhäuser See, 5 Goitsche, 6 Absetzer 1035, 7 Brikettfabrik (briquette works), 8 Muldestausee

Cartography: Dipl.-Ing. T. Herles

Source: LMBV 1999; RP Dessau 1997, 1999; RP Halle 1996, 1998, 1999; Regionaler Planungsverband Westsachsen 1998 b–g, 1999 a,b; 2000 a–d

Fig. 8. The future lake district in the central German coalfield

Flooding the pits will influence the microclimate since the increased evaporation will affect the atmosphere. Higher humidity will result in the immediate vicinity, i.e. within a radius of up to 1 km around large individual lakes or groups of lakes, causing increased mist and fog in the winter and heavy sultriness in the summer. Then again, as the surface area of the lakes expands, it will become windier above the water, which will help reduce this sultriness (cf. Der Deutsche Wetterdienst Brandenburg 2000). Waves and wind will encourage erosion on the shores of the lakes and could endanger the stability of recreation areas near beaches.

2.3
Potential Usage of Mine Lakes

A flooded mining pit can in principle be used to enhance the landscape, for leisure activities and as a water resource (cf. Table 2).

Table 2. Ways of using mine lakes

Mining lakes						
Landscape lakes		Leisure lakes			Lakes as water resources	
Conservation lakes	Lakes for relaxation	Lakes for fishing	Lakes for swimming	Lakes for Water sports	Service water lakes	Reservoir

Source: DVWK 1992

The characteristic feature of a landscape lake is that it is an ecosystem which is largely similar to a natural lake. Priority here is devoted to the development of nature and the countryside. This entails ensuring varied shores and shallow areas, which also improve the aesthetics of the landscape. Lakes of this type are useful for nature conservation and can also be used for 'quiet recreation' (assuming this does not interfere with conservation) not involving the installation of leisure amenities in the immediate vicinity. Conservation lakes are geared towards biotope and species protection, and form regeneration areas or refuges for endangered species (cf. DVWK 1992). Hence the emergence of landscape lakes in the post-mining landscape provides a good opportunity for boosting biodiversity.

A leisure lake is a place where humans can enjoy more active recreational pursuits, including intensive sports. Activities range from walking, cycling and observing nature, through swimming, diving, sailing, fishing and ice-skating, to holiday homes, camping and picnics, to mention but a few. Concrete forms of usage place special demands on the lake in terms of size, depth, water quality and visibility, slope angle, beach formation, footpaths, and utilities and services (supply and waste disposal). Additional criteria also have to be met for more exotic water sports and facilities. If a lake is used for fishing, the fish population has to be geared to the ecological circumstances, and the input of pollution needs to be ruled out in order to protect the lake (cf. ibid.).

A water resource lake can be used to supply service water for industrial and other commercial applications. It can also serve agriculture, the public drinking water supply, and water retention and hence flood protection.

The forms of usage listed in Fig. 9 arise depending on the activities and services conceivable as well as experience in implementing such aims.

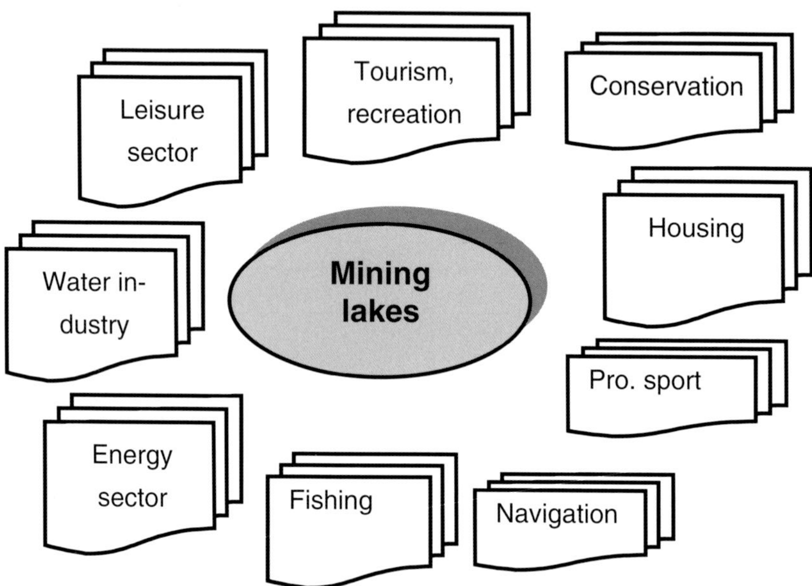

Fig. 9. Specific usage and economic activities at mine lakes

This spectrum shows that shaping the lake landscape (and indeed the post-mining landscape as a whole) is of great importance for the region's structural transformation. The remediation of mining areas means they can be used for a variety of purposes, attracting a number of different sectors.

When planning and creating various services at mine lakes, local circumstances must of course be taken into account. They include factors such as limnological conditions (water surface area, depth, quality, shore composition), the impact capacity of lakes and their surroundings, as well as the type and usage of neighbouring areas. For example, the lake's location (whether it is near large built-up areas or even cities, or close to other lakes) must be taken into account in connection with the usage pressure to be expected and the scope of recreation services to be provided. Attractiveness depends on factors such as whether a lake is adjacent to woodland. Taking into account the local conditions is essential for helping each lake to achieve its own profile. Moreover, plans for individual lakes must be co-ordinated with the amenities and projects at other locations – after all, each lake must also be regarded as a component of the emerging lake district. This enables inefficient repetition to be avoided and also special amenities to be established – which in turn facilitates the creation of a diverse spectrum of usage in the region and promotes the economic sustainability of the individual lakes.

It is equally important to make sure that the specific facilities are geared to demand in terms of breadth and intensity. This spawns the great sociological interest in identifying the ideas and preferences of all the regional actors involved in this process. The actors include institutional decision-makers from the various planning levels and departments, as well as representatives of trade and industry, interest groups, and local inhabitants. Knowledge of their usage intentions and expected intensity considered together with the respective local conditions enables both mutually compatible and conflicting uses to be determined, so that proposals can be made for channelling resulting facilities. The idea is to avoid conflict between different forms of usage at one lake and between neighbouring land uses, as well as to prevent ecological damage and to help conclude sustainable development strategies.

2.4
Tourism – A New Economic and Integration Factor in the Region

Tourism is regarded as one of the boom industries both nationally and internationally. It is an especially attractive sector for regions confronted by fundamental economic and social restructuring. However, the hopes and expectations invested in tourism are often over-ambitious and fail to take into account the current conditions and material requirements. In addition, the feasibility of tourism projects and the related possibilities for creating new jobs within the timescale under consideration are often overvalued. This is exacerbated by the fact that the demands for the sustainable organisation of tourism and the conflicting aims arising are frequently underestimated.

Compared to other areas which are to be developed for tourism, the ravaged mining landscape in the central German coalfield is in a much worse condition. Yet the fact that remediation schemes are to be carried out in order to make the area reusable provides almost unique scope to simultaneously develop both the landscape and options for tourism simultaneously such that they complement each other.

Efforts to create extensive conditions for recreational use in the post-mining landscape are closely connected to the existing scope for landscape development and take into account the high needs of the densely populated region for outdoor recreation and the current shortages in this respect. At the same time they pursue the aim of attracting companies operating in the leisure sector in order to promote the structural transformation of the region. However, merely

implementing the ‘classical’ scheme involving car parks, snack bars and beaches is not by itself sufficient; instead, more exotic attractions are required. However, the creation of such highlights also increases the risk of additional environmental impact. In particular, functions related to nature and the landscape react sensitively to adventure tourism.
Nevertheless, tourism can contribute to sustainable regional development in the follows ways:

- *Ecologically*, if the value of a functionally intact landscape as an important basis for tourism is recognised and the landscape is left largely intact;
- *Economically* as a direct addition to other economic sectors, and indirectly via its numerous interactions and diverse involvement with other economic sectors;
- *Socially*, if the quality of life and original identity of the local population are promoted (cf. Brenner 1999b).

2.4.1 The Ecological Dimension

An ecologically stable natural area is a condition for tourism – yet it is also a factor which the region studied does not yet have. Instead it can only be developed during the course of the next 5–10 years. It is very important to find a balance between extensive and intensive usage forms as well as between usage forms with a regional and a wider attraction, so that the newly forming landscape can be allowed to develop largely free of disturbance and can become ecologically stabilised.
As the development of tourism depends on the region’s ecological condition, it is in the interests of the tourist industry to work in an environmentally friendly manner. The enhanced areas (including lakes) must be managed such that their impact capacity is not strained and that usage conflicts are avoided. The designation of nature reserves enables landscape and species protection. Environmental impacts related to tourism (such as air pollution caused by exhaust fumes, increased volumes of waste, noise disturbance for people and animals, greater surface run-off as a result of land sealing and soil compaction, as well as the pollution of groundwater and surface water) must be minimised by means of the integrated management of the region, which places overall development above individual economic interests.

2.4.2 The Economic Dimension

The development of tourism comprises a major aspect of economic structural transformation since it contributes to both value creation and increased employment. Furthermore, tourism can boost the local economy (e.g. small businesses, retail, transport, medical services, banking) due to higher demand (cf. Brenner 1999b).

With the exception of the spell of 'catastrophe tourism' in the early 1990s, tourism has never been especially important in the central German coalfield. However, this may change as the post-mining landscape takes shape. This has given cause to political decision-makers to place their hopes in the development of local tourism. This is expressed for example in the "Regional Development Forum for Leipzig's Southern Region" founded in January 2001 under the overall control of the Free State of Saxony with the aim of supporting the establishment of tourism by means of conceptual and project-related activities.

High unemployment, a shortage of trainee positions and the hesitant establishment of new business ventures mean that tourism is favoured as an alternative way of generating revenue. However, tourism is a very diverse sector, and so at this point we need to answer the question of which forms should be emphasised for economic reasons, and what forms are suitable for the specific local conditions. According to an analysis from 1995, the growth areas for German tourism are not the typical requirements of holidaymakers (i.e. fortnight package tours) but rather mini-breaks. In areas used for tourism, the proportion of day tourism within total short-term tourism is generally about 75 % (cf. Petermann 1998). Moreover, average spending during short stays is higher than during longer trips. Therefore, the economic effects of short-term tourism (turnover, value creation and yield) are comparatively higher than might at first be assumed from the brief time spent in the area by tourists. As far as overnight accommodation is concerned, classical business tourism as well as trade show and conference tourism are the main segments of demand (cf. Kausch 2000).

The central German coalfield certainly contains the essential conditions for short-term tourism, even if it does not provide all the amenities required. In particular, tourism involving overnight accommodation plays a minor role – with the exception of the cultural (city tourism, festivals) and economic (trade shows, conventions) attraction of district centres like Leipzig and Halle. What predominates is local recreation – a special form of short-term tourism enjoyed during people's free time, i.e. after work and at weekends, and which mainly encompasses outdoor recreational activities. The destinations concerned normally need to be less than an hour away to be feasible. The conditions for this segment of tourism in terms of countryside and infrastructure in the area concerned are comparatively good. In Leipzig's southern region in particular, in recent years the completion of the flooding of a number of mining pits has resulted in new lakes and destinations. Future progress in enhancing the post-mining landscapes will result in the conditions improving throughout Central Germany. The public is very receptive to the creation of conditions for local recreation, especially those which can be classified as sustainable tourism. However, from an economic angle (turnover, taxes, jobs and trainee positions, support for commercial activities), such amenities are not especially profitable and do not contribute significantly to solving employment problems.

Destinations attracting tourists and hence purchasing power from further away are still not the rule. With the exception of cultural projects, the same can be said for seasonal attractions. However, a number of large individual tourism projects are now being planned or have even nearly been completed, kindling hopes that tourists may soon be attracted from further afield. However, the local population tends to be against such large-scale tourism projects, fearing that the improving quality of life and the local environment may be impaired again.

There is a severe lack of co-operation between different tourism companies (strategic alliances), and so the potential synergy effects are not materialised. Combining various leisure-sector activities and increased networking with related economic sectors is essential if the leisure sector is to be fully developed. One important aspect in this respect is the creation of new types of jobs and the development of a service mentality, as the local inhabitants function as multipliers of tourism awareness and are thus an "integrating component of the tourism value creation chain" (OSGV Hessen-Thüringen 2000).

Hardly any attention has been paid to the consequences arising from the transition from development competition to competition for markets once the countryside and infrastructure have been completed. At present, the climate for tourism investments must be judged as somewhat 'unfriendly', which is in particular due to the unsolved legal framework and the lack of corresponding planning bases among local authorities. These aspects requiring urgent attention.

2.4.3 The Social Dimension

The possible positive effects of tourism for increasing the quality of life are undisputed. In addition to decisively helping to establish and secure high-quality environmental and recreational conditions and raising awareness of the importance of keeping nature intact, tourism actually relies on good environmental quality. Furthermore, it is suitable for improving the employment situation and public the supply of goods and services. However, depending on the scope and type of amenities, tourism may also have negative consequences for a region such as an excessive number of tourists, rising transport, overdevelopment of the countryside and dependence on fluctuating demand. These potential hazards underline the necessity of providing a suitable amount of amenities geared to local conditions and demand.

The lack of concrete demand analyses makes identifying the exact target groups rather difficult. Generally speaking, planning and setting up tourist amenities must take into account the interests of potential users in both the target district and the entire catchment area. Priority must be devoted to the values and interests of the local population and other local actors in the area (cf. Leupolt 2000). After all, rural areas are not 'compensating areas' or even 'rest areas' for towns and cities, but are relatively independent areas of living and also economic activity, even if town and countryside are to be regarded in terms of their mutual links and mutual dependence (cf. Teuteberg 1999).

When establishing various types of tourist amenities, the activity-specific modes of behaviour and expectations of users must be considered. Experience from recreation regions in western Germany cannot be transferred willy-nilly to Central Germany as the landscape and socioeconomic conditions are different, and people have different traditions and mentalities regarding leisure activities. For example, spending differences between 'east and west' (largely due to the income gap) and how people spend their free time must be borne in mind.

Knowledge about the interests and behaviour of potential users must be deepened. Special attention needs to be paid to identifying usage preferences and special characteristics with respect to day tourism taking into account modern socio-demographic and socioeconomic development tendencies. The tourism market of the future will largely be shaped by changes in the population and household structure. In this connection, tendencies such as the increase in the proportion of those aged over 60 as a result of declining birth rates and the high migration away from eastern Germany, the increase in households containing just one or two people, the growing level of education, the widening income gap, and the disappearance of traditional working patterns and the growing number of people working outside 'normal' working hours must be emphasised (cf. Petermann 1999). Tourist service-providers must adapt to the changed demand structures and create diverse recreation facilities which take into account the full range of interests. Suitable scope must be granted to both generally popular recreation activities (e.g. swimming and hiking) and less popular interests (such as motor sport).

2.4.4 Facit

Developing the post-mining landscape in Central Germany for tourism could result in the formation of a new regional identity, as long as it is geared towards long-term success. Tourism which is based on the principle of sustainability can act as a connecting link owing to its contribution to securing the protection of nature and the countryside, its contribution to local and regional value creation, its role as an important factor, and its positive effects on the quality of life (cf. Brenner 1999a). Tourism also has an integrating effect in that in the search for unique characteristics it reaches back to the past and in this connection is interested in developing past regional characteristics. The efforts by a region to develop an unmistakable, unique profile by combining mining history with the lake landscape and the new opportunities thereby arising could decisively advance the development of local identity.

Identification with the countryside is also encouraged by its everyday usage. However, as the mining areas were cut off from such usage over a longer period time, the basis is lacking for the development of identification. This shortcoming can be overcome by setting up a "provisional action area" (cf. Barsch et al. 1999) and making the post-mining landscape accessible to the local population while remediation is still underway. The opencast mining tourism which has now been practised for a number of years in various mining landscapes in Central Germany, which takes into account safety regulations and which is conducted by experts, supports this aim. In addition, it also contributes to increasing interest among the local population in the post-mining landscapes, brings the changes to the countryside to life, deepens information about the state of recultivation, and encourages participation in the planning and moulding of the new landscape.

3
Example: Leipzig's Southern Region

3.1
Changing Landscape and Structure – Key Development Prospects

Leipzig's southern region plays an important role within the central German coalfield. It is an area of nearly 500 km^2 containing about 200,000 inhabitants south of the city of Leipzig which – in addition to Leipzig – is bounded by the towns of Kitzscher, Frohburg, Altenburg and Pegau. Although a small part belongs to the Rural District of Altenburg, it mostly belongs to the Rural District of Leipzig.

The mining area known as Leipzig's southern region (which is largely identical with the modern-day planning region) was dominated by lignite mining operated industrially for about 150 years (including 80 years in the form of large-scale opencast mines) and carbon beneficiation. The high volume of mining in Leipzig's southern region (about 50 % of all the lignite mined in Central Germany was extracted here) means that the region underwent extraction of about 35 % (and as much as 70 % at the mining centres). The total surface area removed had reached about 220 km^2 by the late 1990s (cf. Berkner 1998). Some 70 towns and villages were devastated as mining progressed and their approximately 24,000 inhabitants were resettled (cf. Berkner 1993). Other losses included large parts of the original floodplains caused by lowering the groundwater over a large area. Furthermore, severe pollution was caused by emissions from outdated industrial plants used for

briquette production, power generation and coal chemistry. Hence, mining brought about the general deterioration of living conditions.
As a result, large sections of the population migrated away from the region, especially younger and better qualified individuals. In just two years between late-1989 and late 1991, the population in the core district of Leipzig's southern region, the former district of Borna, declined by nearly 6 % (cf. Berkner 1995). Relative consolidation only set in as of 1992 in the rural areas south of Leipzig, with the population slightly increasing as of 1995 (cf. Statistisches Landesamt des Freistaates Sachsen 2000). Nevertheless, the region still features a high population density (see Fig. 10). According to Saxony's district categorisation, the area is regarded as a densely populated area in the vicinity of the regional district centre of Leipzig (cf. Regionaler Planungsverband Westsachsen 1998a). However, one characteristic feature is that Leipzig's southern region is full of opposites – on the one hand there are densely populated areas along the main roads, and on the other empty areas as a result of mining consisting of opencast mines and dumps with the size of up to 40 km^2 (cf. Kabisch and Linke 2000).

Following the changes in eastern Germany resulting from the country's reunification in 1989, in the early 1990s most of the opencast mines and coal beneficiation plants were shut down. Whereas in 1989 nine lignite mines were still in operation, the only one surviving by late-1999 was the mine United Schleenhain comprising Schleenhain, Peres and Groitzscher Dreieck, and where lignite-mining is set to continue for a long time to come.

The far-reaching structural shift in the region calls for a change of economic focus. The development plans for Saxony and West Saxony provide for the development of a territorially balanced, varied economic structure in Leipzig's southern region which is both socially and environmentally sustainable, and aims at the creation of equal living conditions for all local inhabitants (cf. SMU 1994; Regionaler Planungsverband Westsachsen 1998a). Nevertheless, economic development still focuses on traditional core sectors such as mining, power generation and the raw materials industry. By contrast, necessary restructuring processes like the establishment of companies providing special services and working in areas like construction and environmental engineering have only just begun (cf. Ring and Horsch 1998).

Regional planning also attaches great importance to tourism. Leipzig's southern region is to be developed into a "regionally important recreation area" by preserving its distinctive natural and manmade features and taking care not to overstrain the region (cf. Regionaler Planungsverband Westsachsen 1998a).

The basis for this is being formed by the recultivation of disused opencast mines. Activities towards this end were stepped up after 1989. The proportion of land which has been made reusable had been increased from 47 % in the late-1980s (cf. Berkner 1995) to 59 % by the mid-1990s, although the process is subject to much variation throughout the region (cf. Berkner 1998). Hence, even though there is still much recultivation to be done, within just a few years immense changes have taken place and success achieved.

In East Germany (i.e. before 1990), priority was given to restoring agricultural land owing to the high-quality loessy soil. A few mine lakes were created in Leipzig's southern region (which contained few natural lakes) such as Kulkwitzer See, Speicher Borna and Pahnaer See, used for recreation and water management. Nowadays, schemes to reuse Leipzig's southern region (as in the rest of the central German coalfield) are concentrating on flooding mining pits and large-scale reforestation projects (see Fig. 11). Hence the region is also in the grip of landscape change, since remarkable alterations are taking place in land use structures (cf. Table 3).

Fig. 10. The close proximity of built-up areas and mine lakes in the central German coalfield

Fig. 11. Störmthal site at the disused Espenhain mine (Leipzig's southern region), 2001

Table 3. Land use percentages in Leipzig's southern region before lignite mining, now, and following the completion of remediation

Year	1900 [%]	1993 [%]	2050 [%]
Housing, industry and transport	10	15	17–18
Agricultural land	68	48	43–46
Forestry	8	7	18–22
Water	1	4	12–13
Floodplains	13	5	5
Mining	–	21	–

Source: Berkner 1996

About a third of Leipzig's southern region will over the next few decades be characterised by the emerging areas of woodland and water and the remaining floodplain areas. The area of water is set to expand by a factor of eight by 2050, covering an area of 60 km^2 (cf. Kabisch and Linke 2000).

These changes to the landscape (most of which have yet to be carried out) provide suitable conditions for a number of uses, including tourism. Various amenities need to be set up in this region, most of which has a rural character, and which is directly adjacent to the city of Leipzig, which in turn will pave the way for synergetic effects. One important aspect is linking up the remediated post-mining landscape with towns and villages. The supreme aim should be to establish a diversified tourism usage spectrum – which must be in tune with the economic, ecological and social conditions and requirements.

3.2 The Remediation of Water Resources

The remediation of water resources in Leipzig's southern region is the key problem to restoring an intact, largely self-regulating landscape. The goal is to harmonise water qualities in rivers and lakes, which can only be achieved by concepts and schemes focusing on entire catchment areas and transcending administrative boundaries (cf. Berkner 1996).

During the course of the remediation of the opencast mines shut down in 1990, in the example region alone 16 new lakes with a water surface area totalling 52 km^2 and a volume of about 1.16 billion m^3 are to arise. This process will take a number of decades, and will be complete by around the mid-21st century. The timetable for the completion of flooding mining pits can be divided into four distinct stages (cf. Fig. 12).

Between 1995 and 2000, five lakes reached their final state (or – as in the case of Werbener See – reached an intermediate water level which is to be maintained for several years) and have been opened up for general usage. In the following six years, another six pits are to reach the same stage, while between 2010 and 2012 a further three lakes will arise. After a long break, current planning foresees the flooding of two large pits in about 50 years in what is now the United Schleenhain opencast mine (which is expected to be operated until 2040). Hence it can be assumed that in the medium to long term a lake landscape will form in Leipzig's southern region.

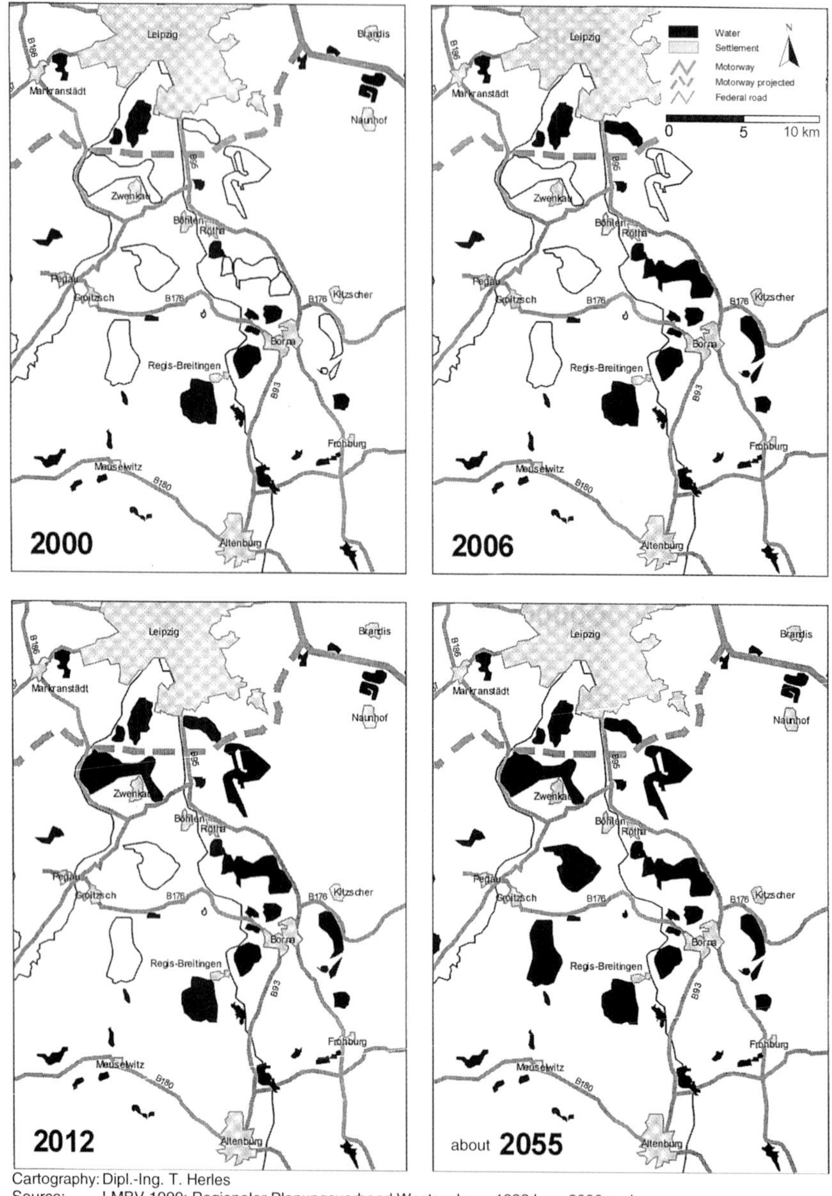

Cartography: Dipl.-Ing. T. Herles
Source: LMBV 1999; Regionaler Planungsverband Westsachsen 1998 b–g, 2000 a–d

Fig. 12. The development of lakes in Leipzig's southern region

The new lakes will vary greatly in size. There are areas will range from between just under 20 hectares to 900 hectares. The time required for the formation depends on factors such as the scope of remediation work necessary (e.g. earthworks for banks and flat areas, slope formation), the volume of water needed for flooding, and the manner of flooding. The water level rises quickly if drainage water from active mines is used, resulting in relatively short flooding times. This also has a positive effect on slope stability and the water quality. In recent years, this method has been used to fill lakes such as Grossstolpener See, Haselbacher See and Cospuden See, and it will also be applied for the new lakes in Markkleeberg and Zwenkau (see Fig. 13 and 14). Hence active mining in Leipzig's southern region (United Schleenhain) and in the immediately adjacent Zeitz–Weissenfels–Hohenmölsen coalfield (Profen mine) is advantageous for the creation of mine lakes.

In contrast to external flooding using groundwater from active mining, allowing mine lakes to form by the natural ascent of the groundwater (as is the case at the future Bockwitzer See) and external flooding using river water (as is to be used at for example the future Pereser See) usually take longer and result in poorer water quality. Acidic pH values result and the water is often polluted. Consequently, a subsequent period of self-cleaning is planned before the water is clean enough to be used for swimming.

3.3 Usage Goals at Mine Lakes

The usage types and goals for the mine lakes already or yet to be flooded in Leipzig's southern region after 1990 are examined below (cf. Table 4) and follow on from all the different ways in which mine lakes can be used (cf. 2.3). The four target categories are also an expression of the different degrees of intensity of water usage. The category "Development of nature and countryside", which includes in particular ecological protection functions as well as selected recreational functions, reflects the lowest intensity, with the highest being reflected by the category "Intensive recreation".

Differentiation between "Intensive recreation" and "Recreation" results from the different breadth of the respective range of leisure and recreation facilities. Whereas "Recreation" mainly comprises opportunities for swimming and 'quiet recreation', "Intensive recreation" covers water sports such as surfing, sailing, canoeing, rowing and diving. The range of opportunities available in turn affects the degree of infrastructural development and the numbers of visitors.

Fig. 13. Swimming at Grossstolpener See near United Schleenhain mine, 2000

Fig. 14. Summer idyll at Haselbacher See after the completion of remediation, 2000

Table 4. Present and future lakes arising in Leipzig's southern region after 1990 (using their current names)

Mining lake	Flooding period	Area of water (ha)	Volume (mill. m^3)	Usage form/aim
Grossstolpener See	1994–1995	30	0.25	*Recreation*, development of nature and countryside
Harthsee	1985–1996	65	6.0	Recreation (recreation usage to be increased as of 2005), *development of nature and countryside*
Haselbacher See	1993–1999	335	23.8	*Intensive recreation*, development of nature and countryside
Werbener See	1990–1999* 2043–2095	79	9.0	*Development of nature and countryside*, recreation
Cospudener See	1993–2000	436	109.0	*Intensive recreation*, development of nature and countryside
Restloch Südkippe	1995–2002	31	1.15	Development of nature and countryside
Restloch Hauptwasserhaltung	1995–2002	18	1.15	Development of nature and countryside
Markkleeberger See	1999–2003	252	60.7	Intensive recreation
Bockwitzer See	1993–2005	170	19.0	*Nature and countryside*, recreation
Hainer See	1999–2006	545	97.8	*Intensive recreation*, development of nature and countryside
Kahnsdorfer See	1998–2006	112	19.9	*Development of nature and countryside*
Zwenkauer See	2005–2010	914	174.0	*Flood protection*, development of nature and countryside, intensive recreation
Störmthaler See	2001–2011	733	158.0	*Development of nature and countryside*, (intensive) recreation
Neukieritzscher See	2009–2012	20	1.6	*Development of nature and countryside*, recreation (limited bathing)
Pereser See	2045–2051	589	138.2	*Development of nature and countryside*, recreation
Luckaer See	1997–2001* 2055–2060	869	339.2	*Development of nature and countryside*, recreation

* Intermediate water level reached

Source: LMBV 1999; Regionaler Planungsverband Westsachsen (Braunkohlenpläne) 1998b–g, 2000a–d

As can be seen from Table 4, mixed usage is practised or intended at most new mine lakes in Leipzig's southern region. For each lake, the main form of usage is shown in bold type. Note that the information covering the lakes to be flooded after 2045 only illustrates the trend of current planning owing to the long time scale involved. Only a few lakes are to be devoted to one form of usage; in these cases multiple usage is ruled out by their small size and hydrological characteristics.

The usage of mine lakes in Leipzig's southern region is mainly geared towards the development of nature and the countryside and the creation of recreation amenities. Only Zwenkauer See is to be primarily used for water management, i.e. flood protection. Nevertheless it, too, will be provided with amenities for quiet and active recreation.

The example of Harthsee, where flooding was completed in 1996, shows that usage intentions may shift within a certain period of time if the external conditions such as the number of available lakes change. Here, the recreation function will be reduced in favour of protection functions for nature and the countryside as soon as Restloch Bockwitz, another nearby mining pit, has been flooded and made usable.

In connection with Werbener See, it should be noted that in 1999 the intermediate water level which is to be maintained for several years was reached. The water level is to start gradually rising again as of 2040, so that the current usage forms can be continued. By contrast, no interim usage is to be established at Luckaer See, where the intermediate water level will be reached next year, until renewed flooding has been carried out in about 50 years' time.

The two main types of usage envisaged at the new mine lakes – the development of nature and the countryside, and usage for recreational purposes – are expressions of the efforts to turn the post-mining landscape into an attractive habitat for both humans and flora and fauna. The objective is to create and maintain high environmental quality. The planned creation of a large number of lakes and the option of the long-term emergence of too many areas of water in the example region and in the central German coalfield reduce the pressure to decide for alternative uses. Instead, these conditions provide enormous development scope since there is plenty of space available for practising diverse extensive and intensive forms of recreation. This range includes both generally popular activities such as swimming and hiking, as well as less common interests in fashionable sports such as motor boats, jet-ski, water skiing and rafting. Such sports are often problematic in terms of landscape sustainability and acceptance, and can therefore not be carried out at every lake. Their organisation calls for suitable protection measures ranging from keeping them a safe distance away from other activities to clear functional separation. In the interests of landscape sustainability and the attractiveness of leisure and recreation areas, a balance must be struck between the natural bases and the development of economically sustainable structures when deciding which forms of usage are to be advocated (cf. Dachverein Mitteldeutsche Straße der Braunkohle e.V. 1999). Moreover, the large scope enables areas of countryside to be protected and retained for usage by future generations. This strategy improves the chances for sustainable regional development.

One key factor regarding the shaping and usability of lakes is land ownership. Lakes and beaches etc. can usually only be kept accessible and be subject to proper conservation measures if they are publicly owned (ibid.).

3.4 Examples of Differentiated Usage of Mine Lakes for Recreation and Tourism

Opportunities for local recreation have perceptibly improved in Leipzig's southern region in recent years. The first lakes are now complete and have been opened up to public usage. Parallel to this, work has begun on developing the necessary infrastructure. For example the network of cycle-tracks and footpaths has been expanded and new bridle-paths have been laid. New maps and information material have been produced. Numerous other projects for shaping the post-mining landscape are now being planned, and some are already nearing completion.

The following examples of planned or completed projects convey the diversity of projects in the former mining areas and in particular the mine lakes geared towards economic development, the development of the countryside, and recreation and tourist amenities. Attention is drawn to special projects and unique features which are suitable for making the emerging landscape attractive for both the local population and also to visitors from other regions. In addition, the selected examples highlight just how complex and complicated the process of making the post-mining landscape usable is.

At the first new mine lakes, traditional intensive and extensive local recreation forms such as swimming, water skiing, sailing, surfing, canoeing, diving, hiking and cycling are especially prominent. The rush to use the lakes opened up to the public, especially obvious during the swimming season of the past few years, underlines the high demand for ways of using these recreational activities.

This is most apparent at Cospudener See, which was formed in the pit left by Cospuden mine shut down in 1992, and which has been used by the public since June 2000. This mine lake is located to the south of Leipzig and west of the adjacent medium-sized town of Markkleeberg. Special needs of the inhabitants of Leipzig and the general conurbation are to be met by attractions such as open-air cinema, rock concerts and similar events. It must be stated that the first few months of public usage were somewhat overshadowed by friction. The intensive usage for recreation hence needs to be co-ordinated with the interests of local residents and conservation requirements.

Cospudener See benefited from healthy finance thanks to its inclusion in EXPO 2000, and so enormous progress was quickly achieved here in shaping the shore areas and establishing an infrastructure suitable for recreational needs. For example, a pier has been built featuring a restaurant, ice cream cafe and a smokehouse for fish, along with a harbour building with a boat shop and moorings, a marina, and a sauna with its own jetty leading right into the lake (see Fig. 15). Other special features include the diving school catering for disabled divers, who can go right up to the edge before being mechanically lifted into the water. One proposal still being discussed is to sink sections of a disused mining excavator to make an attractive destination for driving. Even at a depth of about 20 m, visibility is still 10 m. The southern section of the lake has been reserved for the development of nature and countryside.

The lake can easily be reached by public transport, and is thus kept largely free of individual motorised transport. It can also be reached by a shuttle bus, a network of footpaths and cycle-tracks, and even ferry, and hosts its own bike-hire shops (see Fig. 16).

The types of usage already established at Cospudener See only account for part of the future usage spectrum. Over the next few years, other diverse leisure and recreation services and a suitably sized housing estate are to be built here. The attractiveness of water sports here will be enhanced by linking up Cospudener See to Leipzig's rivers White Elster and Pleisse, allowing access by sports boat, as well as the Zwenkauer See mine lake, which is due to be completed in about 10 years' time (cf. Regionaler Planungsverband Westsachsen 1998c).

Since 1990, a number of attempts have been made to establish an adventure park in Central Germany. One is now being built in the immediate vicinity of Cospudener See and the future Zwenkauer See. "Event Park" is to be built in a number of stages. It is planned to be opened in spring 2002, creating 200 jobs. Annual visitors are expected to total some 800,000. This adventure park for all the family will also include aspects of regional history, including fossilised witnesses of local mining. The highlight of the next stage of development will be the completion of the rafting course in 2003 – by which time the construction of the Leipzig southern tangent (a section of the future A38 motorway) should also be complete. Lobbying is underway to include a separate exit for Event Park so that it can be quickly reached by visitors travelling from further afield.

Interest among the local population in this major tourism project is mixed, ranging from encouragement to scepticism and even rejection. Although the boost it will give the local economy and employment is praised, severe environmental impact is feared. Residents expect their quality of life to be impaired by the intensive land use, high traffic, parking cars and noise.

Fig. 15. The marina at Cospudener See a few weeks after it was opened up to the public in 2000

Fig. 16. The Cospuden – the ferry named after the lake, 2000

Planning for Event Park is integrated into planning of the landscape following the closure of Zwenkau mine (cf. Regionaler Planungsverband Westsachsen 2000c) in 1999. This also goes for the 9-hole golf course to be laid out next to the new marina at Cospudener See (and which may even be expanded into an 18-hole course).

The pit left behind by Zwenkau mine is to be flooded, creating a lake with a water area exceeding 900 ha which is to primarily function as a means of flood protection. However, since flooding occurs rarely, this main function will only play a minor role for a relatively long period. Instead, outside flooding episodes the lake can serve the development of nature and the countryside and be used for recreational purposes. Swimming areas and facilities for boating will be set up (albeit to a limited extent), and in view of the future connection to Cospudener See and Leipzig's rivers a marina may also be built. Plans to preserve the conveyor bridge with a length of 523 m (see Fig. 17) as a major past engineering achievement and the core of an industry and archaeology park as well as the development of housing and a hotel or even a museum are still very much in the balance as their financing is uncertain.

To the east of the former Cospuden and Zwenkau mines – but separated by the B6/B95 (upgraded into a motorway) – is the disused Espenhain mine, which since 1999 has been used to host events for the German Truck-Trail Championships. Parts of this area in its current state are ideal for this sport, the number of venues for which is dwindling in Germany owing to its environmental impact. In 2000 about 12,000 fans and spectators from all over Germany visited this two-day event (see Fig. 18).

Two mine lakes are to be created in the area of this disused mine – Störmthaler See and Markkleeberger See (cf. Regionaler Planungsverband Westsachsen 1998d and 2000d; see Fig. 19 and 20). A lock is to be built so that boats can overcome the anticipated height difference between the two lakes of 4 m. Furthermore, ways of connecting the lake to Leipzig's rivers are currently being examined.

A rowing and canoeing regatta course is to be constructed in the southern section (including the southern shore) of the future Störmthaler See as a water sports venue of national importance. It is to comply with international standards, so that national and international competitions can be held here, and will feature a harbour containing moorings and boathouses, as well as club facilities, a natural grandstand, accommodation, restaurants and bars, and parking. The other shores of the lake will be organised such that usage of the lake is a whole will remain environmentally sustainable. To the west of the mine lake, a conveyor bridge dump with a size of nearly 100 ha will remain, about half of which will gradually be covered with water as flooding continues, resulting in shallow areas. The remaining land will gradually form an island.

Fig. 17. Contemporary witnesses of lignite mining – an overburden conveyor bridge at Zwenkau mine, 2001

Fig. 18. German Truck-Trail Championships at Espenhain mine, 2000

Fig. 19. Remediation work in the area of the future Markkleeberger See, 1998

Fig. 20. Flooding Markkleeberger See, 2001

The future Leipzig southern tangent will run between Störmthaler See and Markkleeberger See (see Fig. 21). Ways are to be sought to integrate the road into the landscape, so that the optical and usage separation effect can be minimised.

One potential danger in connection with both the intended usage of the water and landscape aesthetics is Cröbern Landfill, which was built west of the future Störmthaler See and to the south-east of the future Markkleeberg See on spoil banks, and where public, commercial and industrial waste suitable for disposal with domestic refuse is dumped. It was built back in 1992, when Espenhain mine was still in operation, and whose shutdown and remediation were not yet foreseen.

Another major tourism project designed to attract visitors from further afield is the Underwater Adventure Park at Hainer See (cf. Regionaler Planungsverband Westsachsen 2000b). This unique project in Europe is designed to help up to 600,000 visitors annually to experience the underwater kingdom. Underwater facilities will include a glass tower, adventure areas and hospitality, all connected by pipes. Whether this project will be implemented as planned by summer 2003 largely depends on whether an investor can soon be found. Once the possible underwater area has been flooded, the chances of its materialisation will shrink as this will cause the construction costs to escalate.

The post-mining landscapes and lakes in Leipzig's southern region, which are largely to be intensively used for water sports and tourism, have a counterpart in the three mine lakes at the former Borna East/Bockwitz mine (Bockwitzer See, Restloch Südkippe, Restloch Hauptwasserhaltung) as well as Kahnsdorfer See (yet to be flooded) in the area of the former Witznitz mine, west of the future Hainer See (cf. Regionaler Planungsverband Westsachsen 1998b and 2000b). Priority is to be devoted here to the development of nature and the countryside (see Fig. 22). The usage of geotechnical scope is designed to provide initial locations for succession development, and breeding habitats for birds are to be preserved. Since areas of this type need to be protected, the development of paths will be limited. Selected slope areas, for example on the future Kahnsdorfer See, where remediation will rely totally on nature's powers of self-healing and whose stability cannot be guaranteed, will be closed to the public. However, viewing areas will be set up for the purpose of nature observation.

Planning also provides in a few isolated areas for private and commercial fishing. However, these schemes can only be accomplished in the medium to long term, as the corresponding conditions have yet to be developed. The type of usage (ranging from private angling in a landscape lake to intensive fish farming) must be adapted to the limnological conditions and not conflict with other uses. However, it should be noted that both private and commercial fishing harbour potential for regional value creation and new jobs.

Fig. 21. Filling in the embankment for the planned A38 motorway at Espenhain mine, 1997

Fig. 22. The emergence of landscape lakes at the disused Borna-East mine, 2000

These selected examples of completed and planned projects at mine lakes in Leipzig's southern region show that moulding the local post-mining landscape and the implementation of schemes designed to attract tourists has already led to visible success.

4 Summary

A landscape dominated by lignite mines and industrial plants in much of Central Germany will soon be history. Over the coming decades, the post-mining landscape will be transformed into an attractive lake district. This is not to say that the traditional industries – coal, energy and the chemical industry – do not have a future in this region; indeed, they will remain an important economic factor. However, development concepts and corresponding implementation schemes must take an integrated approach considering natural, economic and social aspects. The usage concepts for the central German coalfield currently being planned observe this requirement, and hence go far beyond the mere continuation of these traditional sectors.

Regarding the shaping and usage of mine lakes, it should be borne in mind that each individual lake is more or less unique in terms of location and specific composition, and that these factors largely determine its new uses. These specific characteristics must be surveyed and developed into regionally unique features in conjunction with other mine lakes and the regional economic and socio-cultural environmental and countryside. This applies in particular to tourism – a comparatively young economic and integration factor in eastern Germany to which high hopes have been attached following the drastic decline of the traditional coal and energy sector. In particular the proximity to the district centres of Leipzig and Halle and the central position within Germany kindle hope that the development of the post-mining landscape for the purposes of recreation and leisure will contribute to the area's further economic development. In this regard, the central German mining region offers almost unique scope and diverse possibilities, as it can be geared towards a completely new range of usage by means of remediation schemes. Providing the remoulded landscape with high environmental quality and making it economically and socially sustainable form the crucial basis for a structural change towards sustainable regional development, during the course of which the quality of life of the local population will be extensively improved in the long-term, the 'soft' location factors will be crucially enhanced, and the emergence of a new regional identity and an enhanced image will be encouraged. Current examples of the region show that the change from a post-mining landscape into an attractive, quasi-natural lake district and a centre of recreation is already underway. However, it should not be forgotten that if these 'recycled' mining areas and lakes are to be used by the public, the links between these new areas and surrounding towns and cities are highly important.

From the viewpoint of the leisure industry, the high spatial density of the mine lakes means that an excessively large proportion of water-based activities can be expected in these local recreation areas. This will not be changed by the mixed usage aimed at in many cases. As a result, competition between the individual locations will become more intense, and it will not be possible for a large number of largely similar types of usage to exist-side-by side in the long term. Moreover, it can also be assumed that usage-intensive activities will worsen the ecological and social impact. Aiming at a model of sustainability which equally considers the natural and cultural conditions for tourism and the interests of both the local population and visitors from further afield may help to defuse this conflict of interests.

The risk of failure in managing the new recreation areas in Central Germany will be reduced by the same extent to which individual projects are integrated into the regional context, decisions and schemes to remould the post-mining landscape are taken by local authorities co-operating with each other instead of individually, and as practical experience in this process increases.

References

Barsch H, Carstensen I., Geldmacher K, Hering F, Jeserigk H, Knothe D, Saupe G, Ziener K (1999) Entwicklung und Gestaltung von Erholungsgebieten in Bergbaufolgelandschaften der Niederlausitz. Potsdamer Geographische Forschungen, Potsdam 17:137

Berkner A (1993) Der Raum Leipzig-Borna-Altenburg – Wege vom ökologisch belasteten Braunkohlenrevier zur Landschaft nach dem Tagebau. Naturforschende Gesellschaft des Osterlandes. Naturwissenschaftliches aus dem Osterlande 3:2–15

Berkner A (1995) Der Braunkohlenbergbau in Mitteldeutschland. Zeitschrift für den Erdkundeunterricht 4:151–162

Berkner A (1996) Die beeinträchtigten Oberflächengewässer des Südraumes Leipzig mit besonderer Berücksichtigung der Pleiße. Hochschule für Technik, Wirtschaft und Kultur (HTWK) Leipzig, Beiträge zu Lehre und Forschung. Zukunft Südraum Leipzig, Sonderheft, 2. Jg., pp 10–21

Berkner A (1998) Naturraum und ausgewählte Geofaktoren im Mitteldeutschen Förderraum – Ausgangszustand, bergbauliche Veränderungen, Zielvorstellungen. In: Pflug W (ed) Braunkohlentagebau und Rekultivierung. Landschaftsökologie-Folgenutzung-Naturschutz. Berlin, Heidelberg, New York, pp 767–779

BMU Bundesministerium für Umwelt, Naturschutz und Reaktorsicherheit (ed) (1994) Umweltpolitik, Wasserwirtschaft in Deutschland, Bonn

Brenner J (1999a) Das Refugienprojekt und die Bedeutung des Tourismus. In: Brenner J, Nehring M, Steierwald M, (eds) Tourismus – ein Beitrag zur wirtschaftlichen und strukturellen Entwicklung für Baden-Württemberg? Ergebnisse des Workshops Nr. IX der Reihe Kommunikation und Verkehr, Arbeitsbericht, Akademie für Technikfolgen-abschätzung in Baden-Württemberg, Stuttgart, pp 3–5

Brenner J (1999b) Urlaubsregion ländlicher Raum? In: Brenner J, Nehring M, Steierwald M (eds) (1999) Tourismus – ein Beitrag zur wirtschaftlichen und strukturellen Entwicklung für Baden-Württemberg? Ergebnisse des Workshops Nr. IX der Reihe Kommunikation und Verkehr, Arbeitsbericht, Akademie für Technikfolgenabschätzung in Baden-Württemberg, Stuttgart, pp 71–77

Dachverein Mitteldeutsche Straße der Braunkohle e.V. (ed) (1999) Straße der Braunkohle, Wasser und Landschaft. Leipzig, p 116

DEBRIV Deutscher Braunkohlen-Industrie-Verein e.V., Bundesverband Braunkohle (eds) (1999) Braunkohle – ein Industriezweig stellt sich vor. Senftenberg, Berlin, 55 S.

Der Deutsche Wetterdienst Brandenburg (ed) (2000) Das amtliche Gutachten des Deutschen Wetterdienstes: Klimatische Verhältnisse der Bergbaufolgelandschaft Südraum Leipzig

DVWK Deutscher Verband für Wasserwirtschaft und Kulturbau e.V. (ed) (1992) Gestaltung und Nutzung von Baggerseen. Regeln zur Wasserwirtschaft, Heft 108, Hamburg, Berlin, p 18

Eissmann L, Campen I (2000) Exkursion in die Bergbau- und Bergbaufolgelandschaft südlich von Leipzig: Geologie, Archäologie, Rekultivierung. Tagungsbegleiter Symposium „Umwelt und Mensch – Langzeitwirkungen und Schlußfolgerungen für die Zukunft. Sächsische Akademie der Wissenschaften zu Leipzig, Leipzig

Grün H-J (1998) Die Umweltbelastungen in den Tagebauen des Geiseltals und Merseburg-Ost. In: Pink R, Schmahl H (eds) Die Mitteldeutsche Braunkohlenregion zwischen Geschichte und Revitalisierung. ARGOS, Wirtschafts- und Regionalmagazin Sachsen, Sachsen-Anhalt, Thüringen. Sonderausgabe Braunkohle. o.O., pp 32–33

Kabisch S, Linke S (1999) Zur sozioökonomischen Relevanz der Nutzung von Bergbaurestsee. In: Horsch H, Messner F, Kabisch S, Rode M (eds) Flußeinzugsgebietsmanagement und Sozioökonomie: Konfliktbewertung und Lösungsansätze. Ergebnisse des Workshops vom 1.u. 2. Juli 1999 am UFZ-Umweltforschungszentrum Leipzig-Halle GmbH, UFZ-Bericht Nr. 30, Leipzig, pp 165–172

Kabisch S, Linke S (2000) Revitalisierung von Gemeinden in der Bergbaufolgelandschaft. Forschung Soziologie vol 97:298

Kausch R (2000) Tourismuswirtschaft in der Stadt Leipzig: Stand und Ausblick. Wirtschaft – Das Journal für die Mitglieder der IHK zu Leipzig 7/8:4–6

Lehmann R (1997) Das Engagement der LMBV für die Revitalisierung. ARGOS, Wirtschafts- und Regionalmagazin Sachsen, Sachsen-Anhalt, Thüringen. Sonderausgabe Braunkohle, o.O., pp 6–10

Leupolt B (2000) Hochwertiger Natur- und Kulturtourismus. Eine Entwicklungsperspektive für Mecklenburg Vorpommern. Zeitschrift für Wirtschaftsgeographie, Jg. 44, 2:113–123

LMBV Lausitzer und Mitteldeutsche Bergbau-Verwaltungsgesellschaft mbH (ed) (1997) Restlochflutung. Gefahrenabwehr, Wiedernutzbarmachung und Normalisierung der wasserwirtschaftlichen Verhältnisse im Lausitzer Revier, Dresden

LMBV Lausitzer und Mitteldeutsche Bergbau-Verwaltungsgesellschaft mbH (ed) (1999) Schaffung von Tagebauseen im mitteldeutschen Bergbaurevier. Berlin, Dresden, Leipzig, 140 pp

Meyer R, Jörissen J, Socher M (1995) Technikfolgen-Abschätzung Grundwasserschutz und Wasserversorgung. vol 2, Berlin, 377 pp

Mibrag Mitteldeutsche Braunkohlengesellschaft mbH (1994) (ed) Die Mitteldeutsche Braunkohlengesellschaft mbH im Überblick und in Stichworten, Theißen

OSGV Ostdeutscher Sparkassen- und Giroverband Hessen-Thüringen (ed) (2000) s-Tourismusbarometer Jahresbericht 2000. Bearbeitung DWIF e.V. an der Univ. München. Berlin, pp 141

Petermann T (1998) Folgen des Tourismus. vol 1, Gesellschaftliche, ökologische und technische Dimensionen (Studien des Büros für Technikfolgen-Abschätzung beim Deutschen Bundestag, 5), Berlin, pp 190

Petermann T (1999) Folgen des Tourismus. Bd. 2, Tourismuspolitik im Zeitalter der Globalisierung (Studien des Büros für Technikfolgen-Abschätzung beim Deutschen Bundestag, 7), Berlin, 270 pp

RP Regierungspräsidium Halle (ed) (1996) Regionales Teilgebietsentwicklungsprogramm für den Planungsraum Profen im Regierungsbezirk Halle. 45 pp

RP Regierungspräsidium Halle (ed) (1998) Regionales Teilgebietsentwicklungsprogramm für den Planungsraum Merseburg (Ost) im Regierungsbezirk Halle, 31 pp

RP Regierungspräsidium Halle (ed) (1999) Regionales Teilgebietsentwicklungsprogramm für den Planungsraum Geiseltal. 51 pp

RP Regierungspräsidium Dessau (ed) (1997) Regionales Teilgebietsentwicklungsprogramm für den Planungsraum Goitzsche. 20 pp

RP Regierungspräsidium Dessau (ed) (1999) Regionales Teilgebietsentwicklungsprogramm für den Planungsraum Gräfenhainichen. 28 pp

Regionaler Planungsverband Westsachsen (ed) (1998a) Regionalplan Westsachsen. Satzungsbeschluss vom 26.06. 1998, 191 pp

Regionaler Planungsverband Westsachsen (ed) (1998b) Braunkohlenplan als Sanierungsrahmenplan Tagebau Borna Ost/Bockwitz. Bautzen, 53 pp

Regionaler Planungsverband Westsachsen (ed) (1998c) Braunkohlenplan als Sanierungsrahmenplan Tagebau Cospuden. Bautzen, 53 pp

Regionaler Planungsverband Westsachsen (ed) (1998d) Braunkohlenplan als Sanierungsrahmenplan für den Tagebau Espenhain. Bautzen, 84 pp

Regionaler Planungsverband Westsachsen (ed) (1998e) Braunkohlenplan als Sanierungsrahmenplan für den Tagebau Haselbach. 48 pp

Regionaler Planungsverband Westsachsen (ed) (1998f) Braunkohlenplan Tagebau Vereinigtes Schleenhain. Bautzen, 94 pp

Regionaler Planungsverband Westsachsen (ed) (1998g) Rohentwurf zur Fortschreibung des Braunkohlenplanes Tagebau Vereinigtes Schleenhain. 21 pp

Regionaler Planungsverband Westsachsen (ed) (1999a) Braunkohlenplan als Sanierungsrahmenplan Tagebau Delitzsch-Südwest/Breitenfeld. Dresden, 87 pp

Regionaler Planungsverband Westsachsen (ed) (1999b) Braunkohlenplan als Sanierungsrahmenplan für den Tagebaubereich Goitsche-Holzweißig-Rösa. 80 pp

Regionaler Planungsverband Westsachsen (ed) (2000a) Braunkohlenplan für den Tagebau Profen. Bautzen, 56 pp

Regionaler Planungsverband Westsachsen (ed) (2000b) Braunkohlenplan als Sanierungsrahmenplan für den Tagebau Witznitz. Bautzen, 57 pp

Regionaler Planungsverband Westsachsen (ed) (2000c) Braunkohlenplan Tagebau Zwenkau. 70 pp

Regionaler Planungsverband Westsachsen (ed) (2000d) Rohentwurf zur Fortschreibung des Braunkohlenplans als Sanierungsrahmenplan Tagebau Espenhain. 18 pp

Ring I, Horsch H (1998) Diversität und regionale Nachhaltigkeit: Entwicklungsperspektiven des Industriestandortes Böhlen. UFZ-Umweltforschungszentrum Leipzig-Halle GmbH, UFZ-Bericht Nr. 8/1998, Leipzig, 102 pp

Schultze M, Klapper H (1997) Möglichkeiten der Beeinflussung der Seewassergüte bei der Fremdflutung von Tagebaurestseen. In: Dachverband Bergbaufolgelandschaft e.V., Vereine und Initiativen in bergbaubetroffenen Regionen, Stiftung Bauhaus Dessau (eds) Jahrbuch BergbauFolgeLandschaft 1997. Leipzig, pp 43–48

SMU Staatsministerium für Umwelt und Landesentwicklung (ed) Landesentwicklungsplan Sachsen (1994) Dresden, 213 pp

SMWA Sächsisches Staatsministerium für Wirtschaft und Arbeit (ed) (1999) Energiebericht Sachsen, Dresden, 58 pp

Statistisches Landesamt des Freistaates Sachsen (2000) (ed) Bevölkerung im Freistaat Sachsen jeweils am 31. Dezember 1989 bis Ende 1999 nach ausgewählten Merkmalen. Kamenz

STUBA Geschäftsstelle des Steuerungs- und Budgetausschusses für die Braunkohlesanierung (ed) (1999) Braunkohlesanierung in den neuen Bundesländern. Berlin, Großenhain, 40 pp

Teuteberg H-J (1999) Stadtnaher Ausflugsverkehr und Tourismus in historischer Perspektive. In: Brenner J, Nehring M, Steierwald M (eds) Tourismus – ein Beitrag zur wirtschaftlichen und strukturellen Entwicklung für Baden-Württemberg? Ergebnisse des Workshops Nr. IX der Reihe Kommunikation und Verkehr, Arbeitsbericht, Akademie für Technikfolgenabschätzung in Baden-Württemberg, Stuttgart, pp 11–44

Thomae M (1996) Die Bedeutung des Braunkohlebergbaus für die Herausbildung industrieller Ballungsräume. In: Dachverband Bergbaufolgelandschaft e.V., Vereine und Initiativen in bergbaubetroffenen Regionen, Stiftung Bauhaus Dessau (eds) Jahrbuch BergbauFolgeLandschaft 1996, Wittenberg, pp 10–21

Umwelt (2001) Kommunale ökologische Briefe, 2:11–12

Woschni E-G, Gebauer H, Schiffer L, Bruckner G, Kranich J, Schuh H (2000) Technikfolgenabschätzung zur nachhaltigen Wasserversorgung und des Gewässerschutzes im Freistaat Sachsen. Abschlussbericht. Sächsische Akademie der Wissenschaften zu Leipzig. Chemnitz, 125 pp

www.braunkohle.de/deutsch/statistiken/beschäftigte.pdf (2001)

Now We Have a Future, People Can Also Speak of the Past – Living Next to an Opencast Mine

Sigrun Kabisch, Sabine Linke

UFZ Centre for Environmental Research Leipzig-Halle, Department of Economy, Sociology and Law, Permoserstr. 15, D-04318 Leipzig, Germany

Everyday life on the edge of an opencast mine means witnessing how landscapes are ravaged on an enormous scale. Huge pieces of equipment such as excavators, spreaders and conveyor bridges are used to dig over entire tracts of land and remove overburden in order to mine the lignite underneath. A high toll is paid for this extraction of raw materials. Entire landscapes full of unique characteristics disappear, the natural balance is impaired, and not even human settlements are safe from destruction. In the central German coalfield alone, by the late 1980s mining had swallowed up some 470 km^2 of the landscape, of which only half had been made reusable. In 1989, 21 opencast mines were still in operation here.

Espenhain opencast mine in Leipzig's southern region (part of the central German coalfield) was finally closed down in 1994. It has an area of about 40 km^2, making it one of the largest former lignite mines. Ever since it was opened in 1937, its gradual expansion destroyed huge areas of fertile agricultural land as well as ecologically valuable floodplains. It also disrupted the local water balance, and 13 towns and villages were razed in its path, evicting some 8,200 inhabitants. Moreover, cleared landscapes with active lignite-mining operations crept into the immediate vicinity of densely populated areas. The impact on everyday living conditions in these adjacent areas (see Fig. 1) and the chances of their revitalisation after the closure of the mines were studied in a sociological research project conducted between 1996 and 1999.[1]

1 Recalling Active Mining

In the case of Espenhain, life on the edge of the mine directly affected the inhabitants of 14 neighbouring towns and villages (see Fig. 2), who were confronted by all the phenomena related to lignite mining with different degrees of intensity and duration. This variation in impact partly depended on the position of the conveyor bridge at any given time and the circular shape of the

Fig. 1. Settlements at the edge of the worked-out opencast mine 1997

mine resulting from this technology. Places where mining operations with preceding overburden removal and subsequent dumping happened to be localised were afflicted by the immediate side-effects such as severe dust and noise pollution. Moreover, paths and roads were severed, which meant that neighbouring villages were suddenly cut off from their surroundings. This prompted the decline of the previous social links and networks of communication. To make matters worse, only a few areas underwent recultivation and became reusable. Residents were never sure if and when recultivation would be carried out, or with what goals. This greatly diminished the quality of life in these communities and ruined their image.

The degree to which adjacent built-up areas were affected also depended on whether they had been classified as "mining protection areas". Classification as such meant that an area had been earmarked for future mining (see Fig. 3). Placing a village under "mining protection" meant that internal investment and building activities were sharply curtailed, with funds to refurbish the building stock and bolster the infrastructure being slashed to next to nothing. These restrictions mostly imposed in the 1970s caused the villages concerned to undergo severe dilapidation. In addition, the living conditions in some areas adjacent to the mine were beset by extraordinarily severe environmental problems owing to the pollutants emitted by the nearby lignite-processing plant. In the words of an inhabitant from one of the neighbouring communities: "When there was an easterly wind we had 'Espenhain air'. Whenever we sat outside to have breakfast, after ten minutes the butter had turned black."

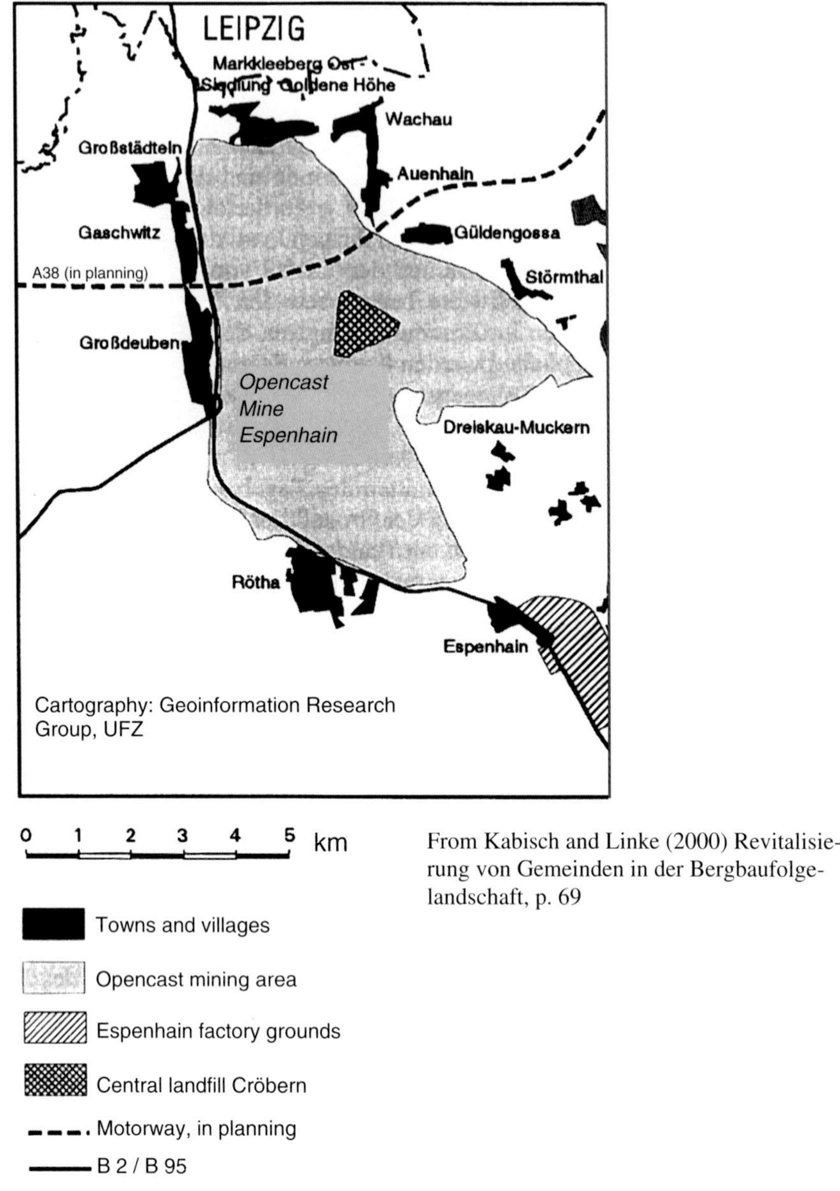

From Kabisch and Linke (2000) Revitalisierung von Gemeinden in der Bergbaufolgelandschaft, p. 69

Fig. 2. The opencast mine Espenhain and its neighbouring towns and villages

As this constellation of conditions meant that the surrounding communities lost all opportunity for development, many of the inhabitants moved away – and the areas' decline was inevitable.

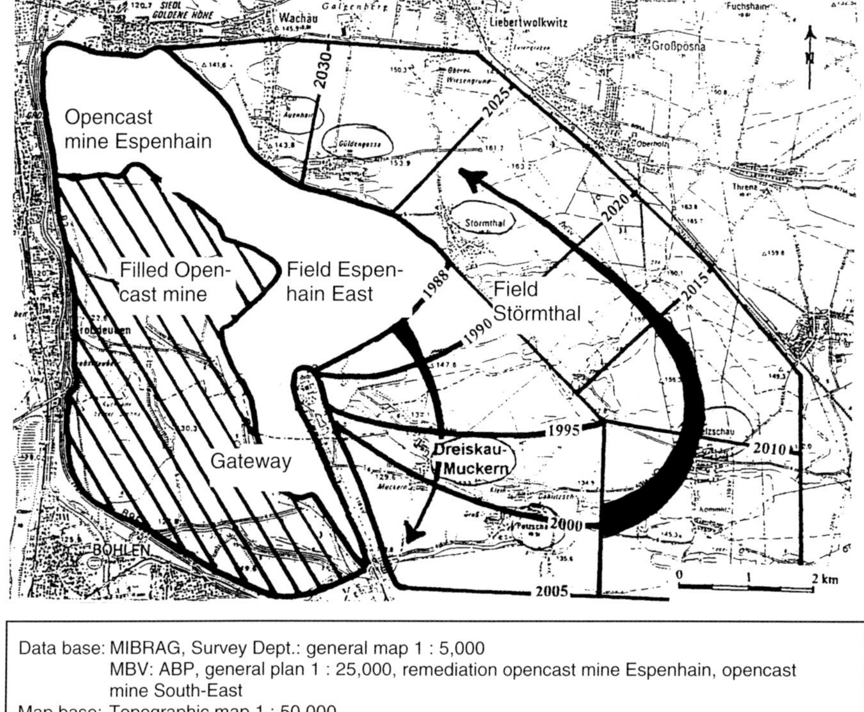

Fig. 3. Planned expansion of the opencast mine Espenhain 1988

2 Changes during Mine Recultivation

The shutdown of the lignite-processing plant in 1990 and the decision to close and recultivate Espenhain mine in 1993 changed the living conditions for the neighbouring communities. Pollution was largely eliminated, and also the "mining protection" status was abolished. This paved the way for the refurbishment and construction of housing, and improvements to the infrastructure. Remarkable progress was made, especially in areas with a high proportion of owner-occupied housing, where the investment of private capital, time and energy combined with state funding programmes led to a dramatic improvement in appearance.

The study findings show that the possibility of using state funding and the inhabitants' willingness to invest their own savings mutually support one another. Although state funding was used to stimulate the revitalisation of the community, by itself it was insufficient; it only led to noticeable improvements in living conditions when applied in conjunction with residents' own commitment. To quote one inhabitant: "I invested all my money in my house – I didn't travel anywhere or do anything else. ... At present we're renovating the basement, and that's not cheap either. You really ought to do it the other way round, but it was only last year that I received the money for refurbishment...." (see Fig. 4 a and b). The positive changes in these areas have been reflected in the arrival of new inhabitants. For long-term residents, this is the best evidence there can be that the negative image of their district has finally been vanquished.

Fig. 4a. Deserted farm in Dreiskau-Muckern 1993

Fig. 4b. People once again see a future for themselves in the villages surrounding the opencast mines. Pictured: a renovated farm in Dreiskau-Muckern 1997

In addition to beneficial local development, the residents in all the areas adjacent to the mine keenly follow the progress made in recultivating the landscape. They hope that this will result in a large number of local recreation areas. They are especially interested in receiving more specific information about the course of recultivation, in particular "details of which parts of the district are to be recultivated, and when". Everyone knows the overall objective, but "more information about the current state of affairs" is desired. Pragmatic questions are asked such as, "When can we residents use these new areas" and "When will the beaches become usable?" Moreover, the way in which the information is communicated is important to interviewees, since "... at the moment everything's too theoretical."

The current changes to the landscape are monitored by local residents with a degree of pride (see Fig. 5). According to one inhabitant, "At the weekend you can see who's got visitors, because they always take them up to the vantage point, although in my opinion soon it won't be very interesting there any more since if you go a few paces towards what used to be Cröbern, you get a view over the whole area. I always send my visitors off on a grand tour – well, I go with them, and you can see everything towards Böhlen and can also look into the mine towards Markkleeberg, where the terraces are still visible, and where the water's already rising. It's lovely. I say we must keep going back there so we can see what's happening."

Fig. 5. People in search of recreation take possession of the opencast mine, 1999

3
Expectations for the Mining Landscape

The basic changes to the landscape will still be able to be observed over the next few years. Following recultivation, the entire district surrounding the communities once next to the mines will be enhanced to form an attractive natural landscape (see Fig. 6).

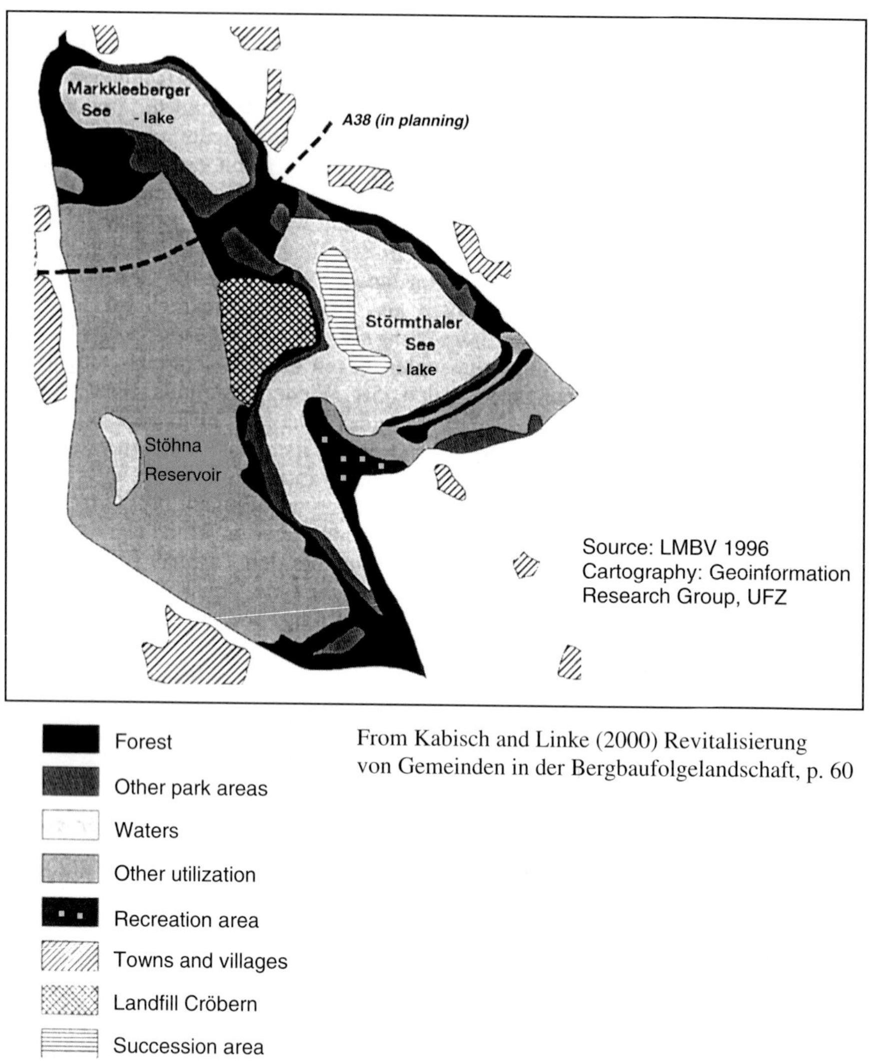

From Kabisch and Linke (2000) Revitalisierung von Gemeinden in der Bergbaufolgelandschaft, p. 60

Fig. 6. Post-landscaping of the former opencast mine Espenhain

Although all the local residents are confident this will materialise, slight differences can be seen between certain groups of communities. One group comprises areas located right on the edge of what used to be Espenhain mine, and whose inhabitants will enjoy direct access to these new recreation districts. These areas also have a relatively attractive, mine-free hinterland, or are embedded in intact surroundings. The second group consists of areas which are separated from the lakes now emerging in the former mine (owing to distance or artificial barriers such as the B2/95 main road). Compared to the latter group, the residents of the former group of villages have higher expectations for the enhancement of the landscape. They emphasise that the recultivation of the former mining district is being accompanied by the emergence of an attractive recreation area on their doorstep. They look forward to a varied landscape and diverse recreational and sports facilities, to surroundings characterised by "lakes, plenty of forest, biotopes and the settlement of indigenous animals species" as well as "cycle-paths, footpaths, swimming beaches and natural areas". Many residents draw express attention to the necessity of considering ecological aspects, with recultivation efforts being directed towards "re-establishing the biological balance in and around mining pits" and the development of "soft tourism" rather than "leisure and fun parks".

Although inhabitants of villages belonging to the second group also look forward to attractive surroundings, they are confronted by obstacles such as the B2/95 and hope for uncomplicated ways of surmounting them.

During the sociological survey, the inhabitants of 14 adjacent communities also expressed how they intend to use the recultivated mining landscape. These intentions are independent of interviewees' distance away from the planned recreational facilities and their foreseeable access. Demand is greatest for quiet, natural recreation (walking, fishing, nature-watching) and for other leisure activities not tied to particular building needs (e.g. swimming, cycling and jogging). Another wish frequently expressed is to open up parts of the mine as an adventure park for children. Although interest in more intensive usage forms (such as jet-skiing and motorboats) is by comparison relatively low, these requirements must also be taken into account.

The process of landscape enhancement which has already begun and the progress to be observed in recultivation are regarded by the local administration and residents alike as catalysts for local development. Owing to foreseeable locational advantages in connection with the planned recreation areas, above all inhabitants of the settlements to the north and south of the mine assume that the results of the recultivation activities will have an enormous positive impact on their villages. In addition to an enhanced landscape and the creation of recreation and sports facilities, they forecast the provision of new building land as well as the further improvement of air quality. In some of these districts, the population also looks forward to new shopping and service facilities and new ways of income for the community as a whole. According to residents, this would in turn raise living quality and encourage other people to move there. The developments over the past few years have shown that these ideas are realistic.

The positive changes in these areas have boosted their image – in the eyes of both local residents and outsiders. Therefore, these areas are expected in future to attract large numbers of day-trippers. This could of course lead to new problems, such as an increase in traffic and "parking on roads which are already too narrow."

4 Regional Problem Areas

When assessing the future chances of these areas, one of the main problems is generally felt to be traffic. Leipzig's planned southern tangent (connecting the A9 and A14 motorways and part of the future A38) goes through the former mine and is generally supported by local residents, who look forward to better links to the regional transport network so they can get to work more quickly in places located some way away from home (see Fig. 6).

Then again, people in some areas here are rather sceptical about these projects. Residents of villages on or adjacent to this new road expect problems such as more noise and traffic. In addition, in the opinion of interviewees from all the areas studied, the planned southern tangent will diminish the quality of the future recreation landscape – by fragmenting it and impairing environmental quality.

According to the interviewees, the landfill at Cröbern (see Fig. 6) built on dumped overburden harbours environmental risks, especially concerning groundwater quality. "Recreation and landfills don't go together," is how one resident sums up the situation. In particular inhabitants of the nearest areas complain about the large number of lorries travelling to and from the landfill and object to the eyesore created by the mountain of rubbish.

Another aspect which makes feelings run high among the local residents is the employment situation. It is already clear to residents that the recultivated mine will hardly produce any new jobs for local inhabitants, and they are keen to see the few existing jobs preserved as long as possible. The resignation and fears of many inhabitants are summed up by the appeal to carry out recultivation very slowly, "so that people don't lose their jobs." Local residents do not expect a job boom during or after recultivation, although the need to take action in this respect is still underlined. However, interviewees do concede that the improvement of the local infrastructure and the establishment of the leisure sector may have a positive, albeit minor, impact on employment.

5 Development Chances for Mining Communities – A Summary

A whole series of positive changes can be observed in all the neighbouring communities ever since Espenhain opencast mine was closed down. These changes are appreciated by local residents, who can still remember all the negative aspects stemming from mining.

In those areas in particular where conditions have fundamentally changed since the early 1990s, a relatively broad spectrum of available development potential can be assumed. Areas where large-scale private and public investment has been carried out are characterised by a mixed, revitalised social structure. This goes especially for those neighbouring districts whose immediate surroundings have been recultivated, and who now have most attractive recreational landscapes on their doorstep.

The developments which have started and the noticeable improvements to everyday living conditions such as the enhanced appearance of villages have gathered their own momentum by for example encouraging further improvements to the building stock and the infrastructure. Positive changes already brought about strengthen ties among the population to their district and also encourage local commitment, thus ultimately enriching social life in the local community. And as the rising living quality can also be perceived from outside, these areas are becoming attractive to outsiders seeking a new home. All the factors considered harbour development potential, guaranteeing these areas a healthy future.

Hence we can see that improving the attractiveness of the landscape just outside Leipzig and the successful revitalisation of communities in the former mining district are two interrelated, mutually strengthening components, which also involve opportunities for a new image for the entire region. This is clearly demonstrated by the study findings achieved by sociologists from the UFZ Centre for Environmental Research Leipzig-Halle. One resident hit the nail on the head by saying: "With hindsight I must say that this is the best thing which could have happened to us – that they decided to put a stop to this independent usage, to quickly repair the [mining pit], and not just to put it off and leave gaping holes in the landscape. And this would have been the case if the pits hadn't been quickly flooded, since the question of usage and thus safety would have been left unsolved.... We won't be overrun here and won't attract any international tourists just because of our new lake, but ... now that we've got a future, the people can also speak of their past. We still have cultural and landscape values which have been preserved. The fact that they're being talked about again will help generate higher acceptance for this region."

[1] The sociological research project "Revitalisation of communities adjacent to mines in the lignite coalfield south of Leipzig – appraisal of the socioeconomic consequences for villages on the edge of Espenhain mine and their future prospects" funded by the German Ministry of Education, Science, Research and Technology (Identifier: 0339674, term: 1 July 1996 to 31 August 1999) focused on 14 areas: Güldengossa, Störmthal, Oelzschau, Pötzschau, Dreiskau-Muckern, Mölbis, Espenhain, Rötha, Großdeuben, Gaschwitz, Großstädteln, East Markkleeberg (with the Goldene Höhe estate), Wachau and Auenhain. A written survey was conducted among all local households, with 1,619 completed and assessable questionnaires being returned.
The scientific results of the project were published in the book of Sigrun Kabisch and Sabine Linke, 2000, Revitalisierung von Gemeinden in der Bergbaufolgelandschaft, Leske+Budrich Opladen, 298 p., ISBN 3-8100-2768-5

Art Projects in Post Mining Landscapes: Visions Became Reality

Ulrich Stottmeister

UFZ Centre for Environmental Research Leipzig-Halle, Department of Remediation Research, Permoserstr. 15, D-04318 Leipzig, Germany

The industrial landscapes of the 19th and 20th century in Central Germany, like those elsewhere, had little or nothing to do with cultural/historical aesthetics. The socialist concept of culture in East Germany concentrated exclusively on the results of the activities of the productive citizens, without concerning itself with aesthetic aspects of the relationship between landscape and industry.

After the political changes of 1989, it is understandable that the further destruction of landscapes by open-cast lignite (brown coal) mining was strongly restricted and that the obsolete industrial plants were closed. Of course, this process was accompanied by devastating social consequences for many residents of the region.

Clearly, the former miners had other concerns than to consider their former workplace from aesthetic points of view at that time. However, an idea born from the plan of the Bauhaus Dessau to develop an "industrial garden" has had an astonishing multiplication effect on the Central German post-mining landscape.

The following examples show how the next generation, having their own perspective towards the former open-cast mines, are remodelling their industrial heritage, taking aesthetic aspects into consideration and creating experience areas for themselves and monuments for their descendants.

These ambitious plans were funded in particular by the LMBV trust (Lausitz-Mitteldeutsche Bergbauverwaltungsgesellschaft), the German federal government and the German states.

1 Example Project: FERROPOLIS

After open-cast mining was discontinued, the scrapping of the excavators, conveyor bridges and overburden spreaders began. These masterpieces of engineering were no longer needed, and were seen purely in terms of megatons of steel and iron which could be recycled with economic benefit rather than being left to corrode.

Martin Brück, a student of the Bauhaus Dessau, developed an astonishing vision, fully worthy of the progressive tradition of this famous school of architecture, in his BA thesis: "FERROPOLIS – City of Steel". To quote from http://www.ferropolis.de:

"Thousands of tons of iron on a left-over peninsula in the closed-down open-cast mining area of Golpa North populate the reclaimed landscape which is to be developed. The excavator town results as a symbol for the technology-fascination and its consequences. No utopia without wounds. The industrial landscapes are the wounds of the utopia of the modern age. In this spirit, FERROPOLIS is an accessible museum of the development of lignite open-cast mining, an artistic sculpture and a landmark."

1.1 Historical Background

In 1958, the Golpa North open-cast mine was opened near the small town of Graefenhainichen in what is now the German state of Saxony-Anhalt. The village of Gremmin, which lay amongst coniferous forests, had to give way to the open-cast mining. Between 1958 and 1991, when the mine was closed, about 70 million tons of raw lignite were mined, serving for electricity production in nearby thermal power stations. About 430 million cubic metres of mining debris were moved. The resulting hole is at present being flooded with water from the river Mulde, and will form a lake with an area of about 580 hectares surrounding the peninsula on which FERROPOLIS is built. The symbolic establishment of the town was carried out on the 14th of December 1995. Funds available for the reclamation of the mine in accordance with the mining regulations (5 m. DM) were used for the project. In this way, the scheduled scrapping of 5 large machines could be prevented and their relocation to FERROPOLIS made possible. A limited company was formed in 1997 to act as the owner, operator and coordinator of the project.

1.2 FERROPOLIS: Art Object, Venue and Leisure Centre

The English stage designer Jonathan Park developed the concept of a Y-shaped open-air arena (Fig. 1) for up to 25,000 visitors. This arena is surrounded by the huge open-cast mining machines which lend in impressive backdrop to events in the arena. Jonathan Park has given easily-remembered names to each of these pieces of equipment: the excavators "Big Wheel" (Fig. 2), "Mosquito" and "Mad Max" together with the overburden spreaders "Medusa" and "Gemini".

The arena is sunk four metres into the ground and will be at the water level of the lake when the flooding is complete. 85,000 cubic metres of earth were moved to build the arena.

Extensive additional building measures were necessary in order to guarantee stability. After the flooding, which started in 1999 and will be finished in 2001, FERROPOLIS will occupy an area of 3 km x 1 km on a peninsula in a 580 hectare lake.

The "steel sculpture" consisting of 5 titanic open-cast mining machines documents an epoch in which the belief in progress and the fascination with technology determined the spirit of the age. The machines are contemporary witnesses as well as being examples of a unique technology. This sculpture shall serve as a memorial for the ecological results of the "landscape consumption". However, it shall also symbolise a new beginning as regards the development of the post-industrial landscape and the new relationships of the local residents to this new landscape.

In the year 2000, FERROPOLIS was included in the EXPO and was the venue for a wide variety of challenging events.

The exhibition of historic rail vehicles is also included in the project.

In this way, FERROPOLIS has become an art object, technology museum, venue and communication centre. However, it is also developing into an important factor for the regional economy.

Fig. 1. Arena under the excavator

Fig. 2. BIG WHEEL in front of the rising lake of FERROPOLIS

Fig. 3. The stage

1.3 Technical Description of the Machines

- BIG WHEEL:
  Shovel-wheel excavator, constructed 1984 (Lauchhammer Heavy Machinery Works), wheel diameter 8.4 m with 14 shovels, weight 1718 t, height 31 m, maximum length 79 m.
- MAD MAX:
  Swinging bucket-ladder excavator, constructed 1962 (Köthen), length 42 m, height 25.7 m, no longer operational.
- MOSQUITO:
  Swinging bucket-ladder excavator, constructed 1941 (R. Wolf Engineering Works, Buckau, Magdeburg), length 53 m, height 21 m, weight 792 t.
- GEMINI:
  Overburden spreader, constructed 1958 (Köthen), total length 125 m, height 30 m, weight 1987 t, length of jib arm 60 m.
- MEDUSA:
  Overburden spreader, constructed 1959 (Köthen), total length 110 m, height 36 m, weight 1132 t, repair crane can be swivelled 360 degrees.

2 Example Project: Remodelled Landscape at Goitzsche near Bitterfeld

In the Central German industrial region, the name Bitterfeld has become a symbol for landscape destruction by open-cast lignite mining as well as for extreme pollution of air and water by an extensive chemical industry, which was a major user of the electrical energy produced by the lignite power stations.

An extensive recultivation programme, begun after the German reunification, included from the beginning not only the design but also the aesthetical aspects of a new landscape. Internationally-known artists were persuaded to participate in the project which, like FERROPOLIS, was included in the EXPO 2000.

2.1 Historical Background

The lignite mining began in 1908 in what was at that time an economically underdeveloped agricultural area near the small town of Bitterfeld in what is now the German state of Saxony-Anhalt. The electricity production in lignite power stations was the basis for the then newly-developed electrochemical processes of chlorine/alkali electrolysis and the corresponding product families based on chlorine chemistry. In other parts of the world (e.g. Buffalo, USA), analogous chemical industries arose with similar environmental problems, although based on hydroelectric power in the example given.

The lignite deposits of the region were mainly to be found under the Goitzsche, 700 hectares of meadow woodland (mixed deciduous forest) situated to the south-east of Bitterfeld. This area, through which the river Mulde flowed, was once an attractive recreation area for the chemistry workers.

About 80 million tons of economically extractable lignite lay under the area, which was therefore made available for mining.

Development of the mines, already planned before the second World War, began in 1949. The various open-cast mines (Holzweißig East, Goitzsche and Holzweißig West) yielded about 430 million tons of raw lignite up until 1991. The villages Paupitzsch, Zöckeritz, Niemegk, Döbern and Seelhausen as well as parts of Petersroda had to give way to the mining schemes. The course of the river Mulde was altered; roads and railway lines were re-routed.

The lignite mining areas to the south and east of Bitterfeld occupied a total of 60 square kilometres.

An unusual feature worth mentioning was the exploitation of the Bitterfeld amber deposits in the lignite open-cast mines, which began in 1970.

2.2
Remodelling of the Goitzsche Landscape

After its closure in the 1970s, the nearby Muldenstein open-cast mine was transformed (by changing the course of the river Mulde) into a reservoir with an area of 6.3 square kilometres, which is now a popular recreation centre.

The area now intended for flooding and landscape design contains several former open-cast mines which will form a region of 9 connected lakes, having a total area of about 23 square kilometres and a shoreline of over 100 kilometres. The largest of these lakes is the Goitzsche lake with an area of 16 square kilometres and a depth of up to 75 metres.

About 550 million Deutschmarks will be spent in the Goitzsche area for remediation and remodelling, provided from funds of the German reunification, the federal government and the German states of Saxony and Saxony-Anhalt.

Extensive work was necessarily throughout the whole area to be flooded in order to ensure stability of the slopes. The dyke-construction for flooding the former open-cast mines was brought into operation in 1998.

The landscape around Bitterfeld must undergo a change of image to become attractive for the local residents and, in order that they are able to remain in the region, also for new industrial developments.

The remodelling of the landscape offers a unique chance to redefine the relationship between mankind and nature by means of the art of landscaping, without forgetting the past. As in FERROPOLIS, this is associated with remembrance of the industrial past, although the essential focus here is on the remodelling of the landscape.

The various landscape architects and landscape artists follow different paths with their work, making use of different local conditions, alluding to former mining situations and aiming to wake emotions.

2.3 Descriptions of Individual Projects

Project: Bitterfeld Waterfront and Water-Level Tower with Pier

Objectives: Urban development of an architectural and artistic connection between the city of Bitterfeld and the new lake;
Architect: Gilles Vexlard, Team Latitude Nord, France; water-level tower Prof. Wolfgang Christ, Weimar (see Fig. 4).

Project: 8 Hills and 49 Conical Heaps

Objective: remembrance of the formerly typical slag-heaps;
Architects: Marc Barbarit and Gilles Bruni, France (see Fig. 5).

Project: AGORA on the Pouch Peninsula

Objective: Park concept with amphitheatre for events and fora;
Architecture: Knoll Ökoplan.

Further landscape art projects refer to former villages which have disappeared, the original course of the river Mulde and other history. They include geological materials of the area in attractive representations and leave cut-up iron parts of mining machines to the mercy of corrosion (see Fig. 6).

The projects are described in detail under:
www.expo2000-sachsen-anhalt.de/content/projekte/goitzsche/landschaftskunst.

The described landscape projects and FERROPOLIS found an astonishing resonance after their opening and especially during the EXPO 2000: they caused a multiplication effect.

Strenuous efforts are now being made to prevent the scheduled demolition of the largest conveyor bridge in Europe (in the open-cast mine at Zwenkau near Leipzig), which was still in use during the year 2000, in order to give it, similarly to FERROPOLIS, a new lease of life as a meeting-place and venue.

Fig. 4. Pier with water-level / outlook tower (height 25 metres)

Fig. 5. Cone landscape

Fig. 6. Hill with iron parts. Interesting colour effects arise under different lighting conditions

References

Leaflet of the Ferropolis Museum and Events GmbH
www.expo2000-sachsen-anhalt.de
www.ferropolis-online.de
www.goitzsche.com

Acknowledgements

The members of the project team would like to thank NATO for the financial support through the linkage grant (ENVIR. LG 960318) and computer network program (CNS 970446). Thanks also go to the contributing editors of the report: Prof. H. Klapper (UFZ), Dr. S. Kabisch (UFZ), Dr. S. Linke, Saxonian Academy of Sciences and Dr. L. Schiffer, Saxonian Academy of Sciences.

We are also grateful to the following companies and individuals for their assistance in this work:

Luscar, Ltd., Edmonton, Alberta, Canada,
R-Princip Most, s.r.o., Czech Republic,
Dr. F. Kennedy, Forestry Commission, Farnham, Surrey, UK,
J.W. Gough, RJB Mining (UK) Ltd., Harworth, S. Yorks., UK,
P.C. Jarman, Celtic Energy Ltd., Aberaman, Aberdare, Wales, UK.

We all would like to thank Ms. Ogarit Uhlmann for her excellent editorial work.

Index

Printing (Computer to Film): Saladruck Berlin
Binding: Stürtz AG, Würzburg